The Aromatic Sextet

The Aromatic Sextet

E. CLAR

Professor of Chemistry
University of Glasgow, Scotland

Foreword by
Robert Robinson

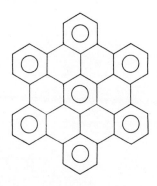

A Wiley–Interscience Publication

JOHN WILEY & SONS

London · NewYork · Sydney · Toronto

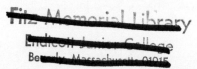

Library of Congress catalog card number: 72-616

ISBN 0 471 15840 2

Printed in Great Britain by William Clowes & Sons Limited
London, Colchester and Beccles

Foreword

Groups of electrons possessing special powers of association or stability in atoms were recognized almost from the beginning of the attempt to apply the electronic theory to the periodic system of the elements. Best known are those that characterize the rare gases of the atmosphere, such as the helium doublet or neon octet.

The idea of the aromatic sextet is analogous in the molecular field. The six electrons indicated by a circle inside the ring are assumed to possess a special stability, so that the type of combination which they represent is not easily disturbed, and should this occur, there will be a strong tendency to restore the original condition. This process of distortion and recovery is noted in many reactions of aromatic substances, for example, in substitutions by means of electrophilic reagents, and in many other types of reaction, of which perhaps the most obvious are the formation and reduction of quinones and their analogues. The characteristic of benzenoid compounds, which is symbolized by the aromatic sextet, is therefore essentially one of stability or resistance to change, and although an organic chemist can recognize the existence of such a property, its explanation in terms of modern theory must be given by the physicist. Certainly the sextet implies very little that can be stated with any degree of certainty about the actual distribution of the electrons in the resting phase of the molecule. Furthermore, aromatic stability is characteristic of aniline, phenol, and pyrrole, as well as of benzene and thiophene, and therefore has no clear relation to reactivity, even when that must involve translocation of the electrons. In all considerations of mechanism of reactions it is preferable to employ the Kekulé formulae and their analogues.

Eric Clar has devoted much attention to the synthesis of suitable polycyclic models, substances designed to test theories of distribution of electrons in the fused nuclei, including the groups of sextets which may be assumed to be present. His extensive spectroscopic work is described in this volume and certainly supports the idea of favourite positions for the sextets, which may be assumed to be present. Some authors have, in effect, stultified the whole theory by using the symbol for the sextet merely to represent potential aromaticity of a cyclic structure. This practice ignores the possibility that the sextet can indeed be taken to represent a relatively stable association of six electrons. If that is the case, then the molecule of naphthalene can only contain one fully fledged sextet. There are not enough electrons available for two. But it is admittedly hard to distinguish between the two nuclei of naphthalene and the choice is apparently between an electron decet or recognition of valence tautomerism, such that the sextet is at one time in one nucleus and the sextet may at different times be in either of the two nuclei. This type of tautomerism can be effected with extremely little change of total energy. A ring which already has four electrons available and is fused to a normal benzenoid nucleus with a sextet, may perhaps be said to have a pseudosextet, that is one that can readily be formed by a tautomeric process. The appropriate symbol can be a small arrow proceeding from the circle, as already suggested by Clar, representing the sextet towards the ring in which the new sextet may be presumed to be formed. Like almost everything connected with the theory of benzenoid compounds, the aromatic sextet is a controversial topic. Eric Clar has made the outstanding contributions in this field by the synthesis of many well chosen examples of polycyclic aromatic substances, by the study of their spectroscopic behaviour, and not least, by rendering available in the present monograph the results of his work in such a form that a conspectus of the stage now reached can be readily obtained. At the lowest estimate it will be seen that he has been able to use the aromatic sextet to assist in collating the data, and although it cannot be said that all problems have been solved and a clear statement uniquely applicable to every case can be made, yet this monograph will leave the reader with the impression that very substantial progress has been made; nevertheless, it is equally clear that difficult problems still call for solution.

If I may mention, what looks like a rather trivial example, it would be the bridge ethenoid bond of phenanthrene, often considered to have special importance in the more substituted derivatives which are carcinogens. The diphenyl nucleus in phenanthrene is doubtless very similar to that in diphenyl itself, with two sextets. They are, however, able to provide a sextet in the middle nucleus if they both make a contribution of electrons to it. Whether or not this can occur is unknown, but the double bond of the phenanthrene bridge is not precisely the same in its properties as the double bond of stilbene. It follows that the two go together in many respects and diverge in some. This is the kind of additional detail which is not the tail to wag the dog. The whole chemistry of polycyclic hydrocarbons bristles with the similar problems and these will doubtless eventually receive some satisfactory solution. Meanwhile, the existence of these secondary difficulties must not blur our vision of the approximate truth of the hypothesis of the aromatic sextet.

In many heterocyclic aromatic structures, unshared electrons of the hetero-atom must be assumed to play a part in the formation of the sextet. The propensity of such electrons to aid the development of a sextet may be increased by salt formation as, for example, in the anions of metallo-pyrrole derivatives, which provide anions more aromatic than the acids from which they are derived. Even more striking are the cases of cyclopentadiene derivatives, since in these cases the formation of a salt on replacement of a proton by a metal cation is the only way in which six electrons can become available for sextet formation.

An aromatic character has been ascribed to a great range of substances, but in many cases the evidence on which this assignment is based is limited to spectroscopic or other physical data, and the chemistry of the substances in question has not been explored in sufficient detail. Some of the large ring polyenes so skilfully synthesized by F. Sondheimer, following theoretical indications of E. Hückel, have been regarded as "aromatic", but the quality of the property is vastly inferior to that characteristic of benzene, thiophene or pyridine. Again, many non-benzenoid aromatic substances are dipoles and the betaine-like character has to be taken into consideration in assessing the total stability of the complex. The situation is somewhat reminiscent of the analogy drawn by Samuel Johnson.

"Sir, a woman's preaching is like a dog's walking on his hind legs. It is not done well; but you are surprised to find it done at all."

6th June, 1972   ROBERT ROBINSON

Contents

CHAPTER 1

Introduction

In 1858 August von Kekulé published a paper[1] in which he clearly connected the carbon atoms to symbolize a bond between these atoms. In this way it was possible to demonstrate that the C atom has four valences which form the bonds. This method also predicted clearly the existence of isomeric compounds in which the C atoms are connected in different ways, as shown below:

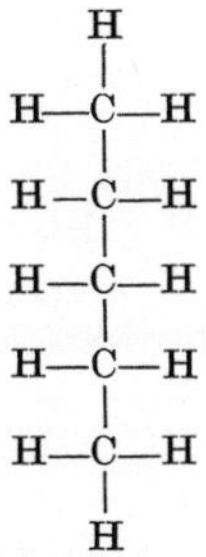

Straight chain

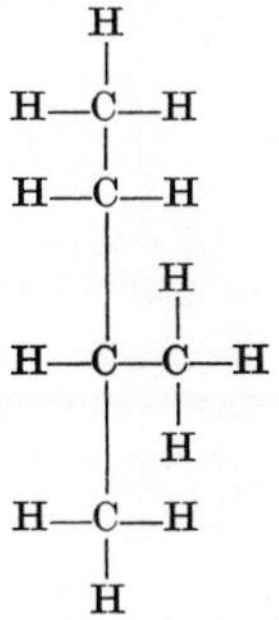

Branched chains

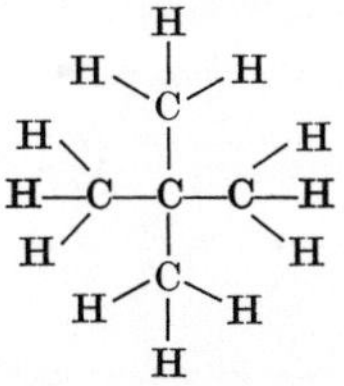

The isomers could not be explained by spherically acting electric forces which were formerly used to formulate inorganic compounds. Such forces would not cause the formation of chains but would result in the collapse of the atoms into a compact

structure with the atoms arranged in the best way for the interaction of coulombic forces.

Kekulé's connection of the atoms had no physical meaning
he even used a mechanical picture involving hooks to demonstrate valences. This did not make the acceptance of his new
structural chemistry easier. However, the facts justified his method
to such a great extent that no other theory had any chance of
competing. But, it should not be forgotten that Butlerow and
Couper had similar ideas.

In his book *Organische Chemie* he did not arrange the four
valences of carbon in a plane, but rather thought them directed
in space. In the first case there would be two isomers **I** and **II**
whilst in the latter case stereoisomerism could be foreseen (**III**
and **IV**).

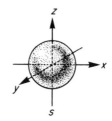

| (I) | (II) | (III) | (IV) |

The optical isomerism resulting from structures **III** and **IV** was
discovered later by Vant Hoff and Pasteur.

Wave mechanics offers a way of giving a physical content to
Kekulé's directed valence, as presented by a line. An s electron
is described as an electronic cloud surrounding the nucleus (Figure
1). A p electron has its electric charge distributed in two such balls

Figure 1. s-Electron orbital

(Figure 2). In contrast to an s bond the p bond derived from a
p electron is directional; therefore three such bonds are possible.

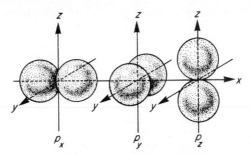

Figure 2.   *p*-Electron orbitals

In order to obtain a tetrahedral bond the process called hybridization must be applied. The electronic configuration of the C atom is $1s^2 2s^2 2p^2$. The promotion to the sp^3 state requires an energy of 60–70 kcal. The transformation of this state into a state with four equal valences pointing into the four corners of a tetrahedron is called hybridization. Figure 3 shows such a hybridized valence electron. It consists of a larger and a smaller lobe. It is obvious that such an electron can form a much stronger bond by overlapping than an *s* electron. The 60–70 kcal needed for the transformation of the $s^2 p^2$ state into the four hybridized orbitals is overcompensated for by the energy obtained by the formation of four bonds, each of which has double the strength of the one obtained from a $2s$ electron.[2]

It is amazing that such a complicated procedure is needed to give a physical content to Kekulé's structural chemistry and to account for the enormous success of his theory which he achieved

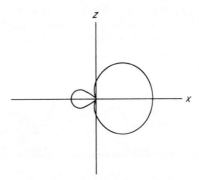

Figure 3.   Tetrahedral orbital

by the uncompromising and strict use of the four valences for carbon and the distinct connection of the C atoms to symbolize the bond between the atoms. This success gives encouragement to proceed with the strict applications of other symbols.

In order to maintain the theory of the four valences of carbon, Kekulé had to symbolize the unsaturated compounds by a double bond. Although this accounts correctly for the four electrons forming the bond it does also mean that the bond consists of two single bonds and must have double the strength of a single bond. This is certainly not true. Numerous facts indicate that there is a single bond of low reactivity and a second bond of much higher reactivity. Thus double bonds give many addition reactions under mild conditions, e.g. the addition of bromine which does not require catalysts or elevated temperatures:

$$\ce{>C=C<} \xrightarrow{\ce{Br2}} \overset{\displaystyle \ce{Br \quad Br}}{\underset{\displaystyle }{\ce{>C-C<}}}$$

The double bond is considerably shorter than the single bond, the lengths being 1.33 and 1.54 Å, respectively. This value cannot be obtained by a combination of two tetrahedrons. The bond energies for the single bond (85 kcal) and the double bond (147 kcal) also show that the two bonds forming the double bond must be of a different nature.

These differences are taken into account by the wave-mechanical interpretation of the double bond. Figure 4a shows the ethylene molecule with the single bond represented by a line and two p electrons having one lobe each above and below the molecular plane before the formation of the π bond. Overlapping of these lobes leads to the structure shown in Figure 4b with the electronic charge of one electron above and one electronic charge below the molecular plane.

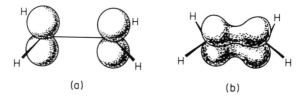

(a)   (b)

Figure 4.   Orbitals in ethylene (a) before and (b) after the formation of a π bond

A simple molecular-orbital calculation (MO method) neglects the electronic correlation of the π electrons, i.e. it does not state their relative position.[3] The ratio of the statistical probabilities of finding both the electrons either above or below or one electron above and one below the molecular plane is $1:1:2$. However, this does not account for the coulombic repulsion between the electrons, which quite clearly favours the latter case to such an extent that it could exclude the first two possibilities. The more classical structure (**V**) represents this case with a σ bond in the centre and one π electron above and one below the molecular plane. If the single bond containing two electrons is symbolized by a full line then it is logical to symbolize a one-electron bond by a dashed line:

(**V**)   (**VI**)

There is also another model for the representation of ethylene. This is the structure with two equal bonds **VI**. These are tetrahedral bonds which are bent. This structure gives the correct bond length for the double bond (1.33 Å) if the arc of these bonds is assumed to have the same length as the single bond (1.54 Å). Even the angle between the double and single bonds is correctly predicted to be $125.27°$.[4] This is just one example in which wrong premises produce correct results in theoretical organic chemistry. However, nuclear magnetic resonance (n.m.r.) measurements exclude this formula, because the magnetic information transmitted through this bond as measured by the coupling constant (J) should have double the value of a single bond. This is not the case (see p. 34).

References

1. A. Kekulé, *Liebigs Ann.*, **106**, 129 (1858); *Lehrbuch der Organischen Chemie*, Verlag Ferdinand Enke, 1861, 1, 114, 157, 172, 183; *Zeit. Chem. (NF)*, **3**, 214 (1867).
2. L. Pauling, *J. Amer. Chem. Soc.*, **53**, 1367 (1931); J. C. Slater, *Phys. Rev.*, **37**, 481 (1931); L. Pauling, *The Nature of the Chemical Bond*, Cornell University Press, New York, 1960, p. 108.
3. M. J. S. Dewar and H. N. Schmeising, *Tetrahedron*, **11**, 96 (1960).
4. L. Pauling, *The Nature of the Chemical Bond*, Cornell University Press, New York, 1960, p. 137.

CHAPTER 2

The Structure of Benzene

In 1865 Kekulé[1] published his hexagon structure of benzene (**I**). It brought an insight into the chemistry of "aromatic substances" and it enabled the principle of the four valences of carbon to be applied to these cases. This formula contains three double bonds and therefore accounts correctly for the number of electrons in the system. However, as stated in the preceding chapter, the representation of a double bond by two single bonds does not properly describe the properties of a double bond. One would expect that these difficulties would become even worse with three double bonds in a cyclic arrangement. Apart from this, other difficulties quite soon became apparent.

(**I**)   (**II**)   (**III**)

There should be two isomeric *ortho*-substituted derivatives (**II** and **III**), the first having one double bond between the substituents R and the second none. These could not be found. Kekulé took account of this fact by the introduction of his oscillation hypothesis which assumes that the positions of the double bonds are continuously changing.

The same difficulty arises if rings are fused to the benzene ring (e.g. **IV** and **V**). Thus ring 1 in formula **IV** must be different from

(**IV**)   (**V**)

6

ring 2. The same is true for formula **V** in which the double bonds in the central ring have changed places. Any difference between rings 1 and 2 cannot be found. The transition from benzene to naphthalene and anthracene is quite uniform as measured by the absorption spectra. This is called the uniform annellation effect.[2]

There are more facts incompatible with the presence of three true double bonds in benzene. The sum of the bond energies of the six C—H bonds plus three C—C and three C=C bonds gives 1286 kcal. However, from the heat of combustion a value of 1323 kcal is obtained. The difference of 37 kcal could therefore result from the interaction of the two Kekulé structures. It is called resonance energy in the valence-bond (VB) theory.[3]

The heat of hydrogenation can also be used to measure the interaction of the double-bond structures. The heat of hydrogenation of cyclohexene to cyclohexane is 28.39 kcal. If there were

three non-interacting double bonds in benzene three times this value should be obtained. However, the observed heat of hydrogenation of benzene is much smaller (49.80 kcal). The difference

of 35.97 kcal comes very near to the value obtained from the heat of combustion and must result from "aromatic interaction".

That such an interaction must take place is also shown by the X-ray analysis of benzene which results in a regular hexagon with a side length of 1.39 Å. This excludes alternating single (1.54 Å) and double bonds (1.33 Å). If the aromatic bonds were 50 per cent double bond in character one would expect a bond length of 1.435 Å. The compression of 0.045 Å for one bond is the result of the energy obtained by the aromatic interaction or delocalization of the π electrons.

The MO theory uses the same molecular orbital as for the double bond. The two single Kekulé structures of benzene are

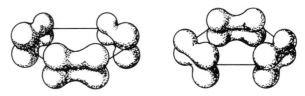

Figure 5.   Molecular orbital of the three double bonds in benzene

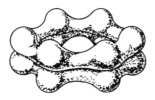

Figure 6.   Molecular orbital of delocalized double bonds in benzene

represented in Figure 5. The interaction leads to the molecular orbital as shown in Figure 6. This can be considered as a rotation.[4]

Hückel's MO theory[4] predicts two occupied orbitals for benzene, the lower being occupied by two π electrons with opposed spins. These are not free to travel round the ring. The next highest orbital contains two pairs of π electrons, which can pass round the ring pairwise in opposite directions. Therefore the system is diamagnetic and anisotropic. This can be represented by the structure in Figure 6.

The energy resulting from the delocalization of the π electrons is measured in terms of an energy unit β. This amounts to 18–20 kcal and it is the energy obtained from one of the two π electrons forming a double bond. A Kekulé structure having three localized double bonds should have the energy of 6β. The MO treatment predicts an energy of 8β for the benzene molecule with delocalized double bonds. The excess energy (2β) results, therefore, from the delocalization. The double bonds of the Kekulé structures disappear in the ground state and are replaced by a new bond type with rather more than 50 per cent double-bond character.

In Hückel's MO theory great importance is attached to the group of six (2 + 4) π electrons which gives stability to the ring system that would otherwise have an unsaturated character. Before the MO treatment Bamberger[5] had already explained the stability of furan (**VI**), thiophene (**VII**) and pyrrole (**VIII**), by the assumption of "six potential valences", which Hückel later interpreted as a group of six π electrons in which the lone pair of electrons of the heteroatom participates. However, these systems have not the symmetry of benzene and there must be some limitation of the delocalization which mainly concerns the lone pair of the heteroatoms.

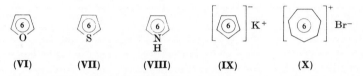

| (VI) | (VII) | (VIII) | (IX) | (X) |

The above group with six delocalized π electrons can be extended to the ionic compounds **IX** and **X**. The K atom in cyclopentadienyl potassium (**IX**) contributes one electron to the group of six. In cycloheptatrienyl bromide (**X**) the Br atom removes one electron from the seven π electrons of the ring thus forming the stable group of six.

Hückel's theory predicts particular stability for all cyclic systems which have $(2 + 4n)$ π electrons, n being an integer. The main series of this kind is formed by the acenes, i.e. linearly annellated benzene rings: benzene, naphthalene, anthracene, tetracene, pentacene, hexacene and heptacene. These have 6, 10, 14, 18, 22, 26 and 30 C atoms and π electrons, respectively. The higher members of this series are deeply coloured, highly reactive and very unstable. Hückel's rule does not discriminate between acenes and the angular annellated phenes which form the same series with $(2 + 4n)$ π electrons but show a much greater stability. Moreover, it does not fit the series of the most stable hydrocarbons which are fully benzenoid and have multiples of six π electrons. It is therefore obvious that Hückel's rule must be strictly limited to monocyclic systems (see Chapters 6 and 7).

Another method of giving a quantitative description of the aromatic bond is the valence-bond (VB) method which was introduced by Pauling.[3] It uses classical structures: the two Kekulé structures **XI** and **XII**. The interaction between these structures is called resonance. It results in a ground state which cannot be represented by any Kekulé structure but by a resonance hybrid which has a lower energy than a single Kekulé structure. This

ground state does not contain double bonds. Besides the Kekulé structures the three Dewar structures **XIII**, **XIV** and **XV** have to be considered. These are supposed to be less stable because of the long *para* bond. They contribute less to the ground state of benzene. The VB method gives the Kekulé structures a contribution of 78 per cent and the Dewar structures 22 per cent. The resonance energy is calculated in units of α, the interaction energy of the two p electrons which transforms the single bond into the double bond.

The resonance energy calculated from the two Kekulé structures is 0.9α. This value is increased to 1.11α if the three Dewar structures are included.

Dewar benzene and a number of its derivatives were synthesized later.[6] However, their independent existence cannot be used as an argument against the resonance theory. Dewar benzene (**XVI**) is not planar and the *para* bond is a single σ bond. Only the existence of a planar compound with a single *para* π bond would contradict the theory.

As with the MO theory the difficulties begin with the annellated systems. An extension of the VB theory to naphthalene uses the three Kekulé structures **XVII**, **XVIII** and **XIX**.

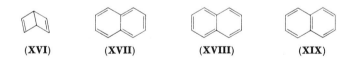

For anthracene there will be five Kekulé structures. In addition to this, the Dewar structures have to be taken into account. The number of structures to be considered is thus rapidly increasing and the calculation is therefore enormously more complicated.

There can be no doubt that mirror-like structures **XVII** and **XVIII** have the same energy. However, this is not so certain for the structure **XIX** which has a double bond between the two rings. This problem will be dealt with later using n.m.r. measurements.

References

1. A. Kekulé, *Bull. Soc. Chim. (NS)*, **3**, 98 (1865); *Zeit. Chem. (NF)*, **1**, 277 (1865); *Bull. Acad. Roy. Belg.*, [2] **19**, 551 (1865); *Lehrbuch der Organischen Chemie*, Verlag Ferdinand Enke, Erlangen II, 1866, p. 493, 514; *Liebigs Ann.*, **162**, 77 (1872); **221**, 230 (1883); A. Kekulé and O. Strecker, *Liebigs Ann.*, **223**, 170 (1884); A. Kekulé, *Ber. Deut. Chem. Ges.*, **2**, 362 (1869).

2. E. Clar, *Polycyclic Hydrocarbons*, *I*, Academic Press, London, 1964, p. 32.

3. L. Pauling, *J. Chem. Phys.*, **1**, 280 (1933); L. Pauling and G. W. Wheland, *J. Chem. Phys.*, **1**, 362 (1933); L. Pauling and J. Sherman, *J. Chem. Phys.*, **1**, 606, 679 (1933); L. Pauling, *The Nature of the Chemical Bond*, Cornell University Press, New York, 1960, p. 198.

4. E. Hückel, *Z. Physik*, **70**, 204 (1931); **72**, 310 (1931); **76**, 628 (1932); **83**, 632 (1933); *Grundzüge der Theorie ungesättigter and aromatischer Verbindungen*, Verlag Chemie, Berlin, 1938, p. 71.

5. E. Bamberger, *Ber. Deut. Chem. Ges.*, **24**, 1758 (1891); *Liebigs Ann.*, 273, 373 (1893).

6. E. E. van Tamelen and S. P. Pappas, *J. Amer. Chem. Soc.*, **85**, 3297 (1963); E. E. van Tamelen, S. P. Pappas and K. L. Kink, *J. Amer. Chem. Soc.*, **93**, 6092 (1971).

CHAPTER 3

Robinson's Aromatic Sextet

Many years before the wave-mechanical treatment of the aromatic bond Robinson[1] introduced a circle inside the hexagon to symbolize the six π electrons of benzene. This symbol is intended to stand for the mobility of the π electrons in benzene and for the stability of benzenoid compounds. This concept can easily be extended to furan, thiophene, pyrrole and other heterocyclic systems in agreement with their chemical and physical behaviour. What was a "glimpse of the obvious" to the originator of the idea in 1925 appears today as one of the rare manifestations of chemical instinct of the Kekulé style.

The advantage of applying the right symbol is to avoid a detailed discussion of the fine structure of benzene and to be able to wait for the fruits which the logical application of the sextet will bring. Thus the circle must be applied in the same logical way as Kekulé used the line connecting the atoms, which proved to be the very basis of structural chemistry. Keeping this in mind the circle cannot be used to indicate aromaticity or potential aromatic rings. Therefore formulae **I**, **II**, **III** and **IV** for naphthalene, anthracene, phenanthrene and perylene, respectively, are highly objectionable.[2]

The circle in these formulae stands for six and four π electrons in naphthalene (**I**), for six, two and six (or six, four and four) π elec-

(I)　　　　**(II)**　　　　　　　**(III)**　　　　　**(IV)**

trons in anthracene (**II**) and six, two and six π electrons in phenanthrene (**III**) and even symbolizes no π electrons in the centre ring
of perylene (**IV**). No structural chemistry can be based on this use
of the circle. If the aromatic sextet invariably symbolizes six π
electrons in a monocyclic system, typical aromatic or, better,
benzenoid stability can be represented. There is no reason to call
the sextet the benzenoid sextet, because if the sextet is shared
among more rings then the benzene-like character is diluted to
such an extent that rather unstable systems are formed which
nevertheless are aromatic.

There can be only one sextet in naphthalene (**V** and **Va**). In
order to make the two formulae symmetric one must assume that
two π electrons of the sextet can migrate from one ring to the other.
This is symbolized by an arrow in formula **VI**. The two double bonds
in this formula do not indicate more than that there are four π
electrons in these rings. In fact it does not contain two true double
bonds just as benzene does not contain three double bonds (see
Chapter 10). The symbol of the double bond in rings in this
sense is at first retained to make the transition from the conventional formulae easier. Instead of the double bonds one could just
write the figure 4 into the ring (**VII**).

| (V) | (Va) | (VI) | (VII) |

There is only one sextet in anthracene (**VIII** and **VIIIa**) which can
be in one of the two side rings or the middle ring. Formula **IX** is a
simplified expression for this.

| (VIII) | (VIIIa) | (IX) |

It is obvious that if in a higher acene, as this series is called, one
sextet is shared among several rings, this must necessarily lead
to a gradual loss of benzenoid character. In fact, the reactivity
increases rapidly in the acene series: benzene, naphthalene,
anthracene, tetracene, pentacene, hexacene and heptacene.

Naphthalene does not show the high stability of benzene and is sulphonated by concentrated sulphuric acid at room temperature. Anthracene is still more reactive and is easily oxidized in the middle ring to anthraquinone. Tetracene is orange and gives readily a photo-oxide with air and light in the middle ring. The photo-oxidation becomes such a danger to the preparation of the higher acenes that the violet pentacene and the green hexacene must be handled under nitrogen. Finally, the dark-green heptacene is so unstable that it has never been obtained in a pure state.[3] The movement of the sextet through the rings in heptacene is shown below:

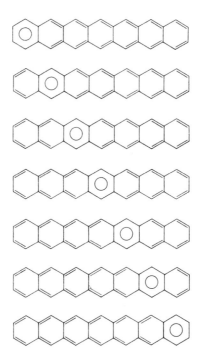

It is obvious that with an infinite number of rings the system becomes a cyclic polyene in which one sextet is not enough to give some degree of stability.

Whilst aromatic rings which share one sextet can have any degree of stability ranging from benzene to free diradicals, an "inherent" sextet in only one ring must approximately provide the stability of benzene. It becomes necessary to discriminate between aromatic and benzenoid rings, the latter always having the highest stability.

This can be shown by a comparison of anthracene (**IX**) and phenanthrene (**X**). Anthracene has only one sextet which is shared between three rings.

(**IX**)　　　　(**X**)　　　　(**XI**)

There is one angular ring in phenanthrene (**X**) which has three double bonds and by definition must be written with a sextet as in **XI**. Therefore the aromatic energy of phenanthrene is 7–12 kcal greater than that of anthracene.[4] However, this statement applies only to the whole molecule. Whilst the rings marked with the sextet show benzene-like stability the central ring has a fixed double bond. In contrast to the double bonds written in naphthalene (see p. 13, formulae **V**, **Va** and **VI**), this is a true double bond. Its localization is the result of the formulation with two sextets. This double bond is indeed as reactive as any olefinic double bond and it adds bromine without a catalyst to form the relatively stable dibromide (**XII**). This justifies fully the formulation with two inherent sextets and makes other structures like **XIII** or **XIV** unlikely:

(**XII**)　　　　(**XIII**)　　　　(**XIV**)

Formula **XIV** has only one sextet which is an inherent sextet, whilst formula **XI** has two inherent sextets and in the central ring an induced sextet. This is formed by the two π electrons of the fixed double bond and two pairs of π electrons each transferred from the neighbouring ring as indicated by the arrows. This induced sextet must not be symbolized by a circle because such a misleading formula would show eighteen electrons whilst formula **XI** has the correct number of fourteen π electrons.

Before further comparisons of this kind are carried out it is necessary to apply a correct scale for colour and reactivity. This is provided by the absorption spectra.

References

1. T. W. Armitt and R. Robinson, *J. Chem. Soc.*, **1925**, 1604.
2. R. Robinson, *Aromaticity, Intern. Symposium Sheffield 1966*, p. 47.
3. E. Clar and B. Boggiano, *J. Chem. Soc.*, **1957**, 2683.
4. J. W. Richardson and J. S. Parks, *J. Amer. Chem. Soc.*, **61**, 3543 (1939); A. Magnus, H. Hartmann and F. Becker, *Z. Physik. Chem.*, **197**, 75 (1951).

CHAPTER 4

Electronic Spectra and Classification of Absorption Bands

The Electronic Spectra

An absorption band is the result of an electronic transition from the ground level to a higher level. This transition is superimposed by nuclear vibration in the upper state. The first absorption band is therefore followed by a group of bands, as shown in Figure 7. The reversal of this process is observed in fluorescence spectra. Here the electronic transition is superimposed by nuclear vibrations in the ground state.

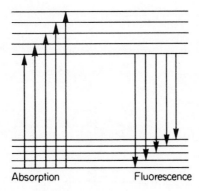

Absorption   Fluorescence

Figure 7.   Term scheme for absorption and fluorescence

Figure 8 shows a comparison of the absorption spectra of benzene, phenanthrene, pentaphene and heptaphene. The different groups marked α, β, p (*para*) retain their characteristic features whilst a strong shift towards the red is observed. This is called the annellation effect.[1] It demonstrates that the two branches are in aromatic

17

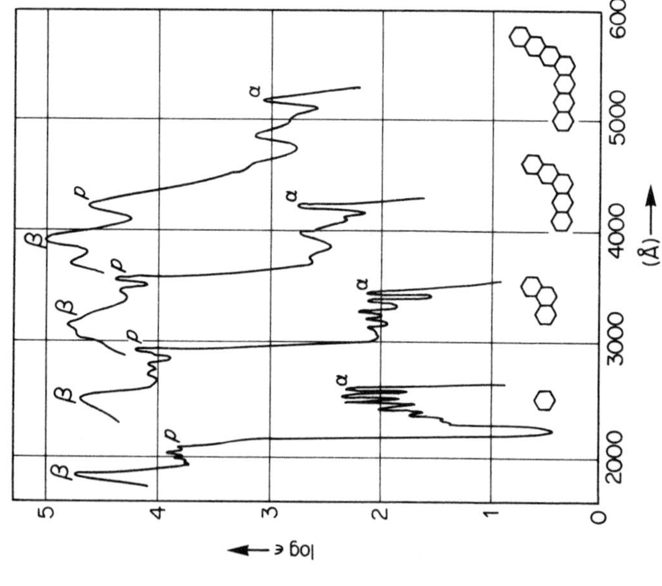

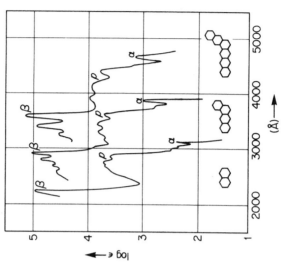

Figure 8.   Absorption spectra in the phene series

Figure 9.   Absorption spectra of the asymmetric phene series

conjugation, each ring producing a shift towards the red which is about constant in $\sqrt{\text{Å}}$. This rule is valid as long as the systems contain only two sextets with the exception of the first member of the series, benzene.

If the annellation of the two rings does not take place in a symmetric way as in the above phene series a similar asymmetric phene series consisting of naphthalene, tetraphene and hexaphene is obtained (Figure 9). These hydrocarbons also contain only two sextets with the exception of the first member, naphthalene. The two branches which are in aromatic conjugation again produce constant shifts in units $\sqrt{\text{Å}}$. However, from the beginning of the series the p bands have shifted more than the α and β bands. This causes a partial superimposition of the α bands of which only one band is left visible. It will be shown later (p. 20) that the migration of the p bands is essentially dependent on the longest

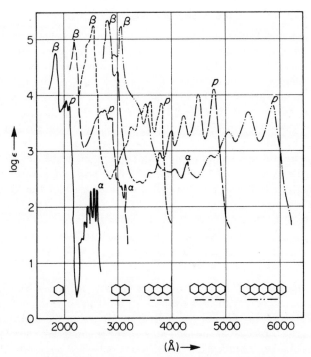

Figure 10. Absorption spectra of the acene series

branch, whilst the shift of the α and β bands responds equally to the linear annellation of rings to any of the two branches. The ratio of the wavelengths of the α to the β bands remains about constant at 1 : 1.35.

If the asymmetry in the phene series goes to the extreme then the linear acene series is formed. Figure 10 demonstrates that the p bands shift more to the red with annellation than the α and β bands. Thus the α bands become completely hidden by the p bands in anthracene and tetracene. They reappear faintly in the minimum of the spectrum of pentacene. Since only one branch is extended with annellation the p bands show the strongest shift to the red. However, their shift is again constant in $\sqrt{\text{Å}}$. This empirical rule is of great value if the degree or the absence of aromatic conjugation is to be measured.

The Classification of the Absorption Bands

In the acene series the p bands have an intensity of about $\log \epsilon = 4$ which changes very little with annellation. This increases very much if acenes are condensed vertically, for instance in the *rylene* series: naphthalene, terrylene and quaterrylene (see pp. 83, 93). The p bands are shifted towards the red at low temperature. Another red shift of about 900 cm^{-1} is observed in passing from the gaseous state into solution in hexane, heptane or alcohol. A further shift of 300 cm^{-1} is recorded for the solution in benzene.[2]

There are numerous indications that the p bands are correlated to a localization of two π electrons in the *para* position, because the reactivity in this position is very much increased in the higher acenes:

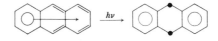

This localization is favoured, because it produces another sextet, i.e. it stabilizes the excited state. The intensity of the p bands is greatest if polarized light is irradiated in the molecular plane in the direction of the long axis a. The p bands can therefore also be called L_a bands. This transition is polarized perpendicular to the long axis.

The α and β bands are polarized perpendicular to the short axis b of the molecule. This means that light polarized in the molecular plane is most strongly absorbed in the direction of the short axis b. The α and β bands have been termed L_b and B_b bands respectively.[3] This classification of bands based on the molecular axes shows deviations in angular hydrocarbons and becomes completely misleading in condensed hydrocarbons of the perylene type.[4] It is therefore better to retain the more general terms α, β and p bands.

The α bands are always visible in the series of the phenes. Their intensities are usually in the range log ϵ = 2–3. However, they can reach higher intensities when polycyclic systems contain five-membered rings as in the fluorenes, biphenylenes and fluoranthenes. The α bands shift very little with falling temperature to shorter wavelengths. The transition from the gaseous state into solution in hexane or alcohol brings a red shift of about 250 cm^{-1}. This shift is dependent on the intensity and is bigger for higher intensities.

The β bands are the most suitable bands for the comparison of annellation effects. They shift towards the red with linear annellation of the two branches. They are also the most intense bands in the spectrum. It is therefore not possible that conclusions drawn from their position or appearance are based on the presence of small amounts of impurities which could very well be the case when using the α bands. The β bands show a shift of about 900 cm^{-1} to longer wavelength when passing from the gaseous state into solution in hexane or alcohol. Another shift of about 300 cm^{-1} is observed when passing into solution in benzene. They show a considerable red shift at low temperature. The β bands are never superimposed by other bands and can always be observed, independent of the symmetry of the molecule. Hydrocarbons of the phene or acene series with the same number of rings exhibit β bands at the same wavelength. Only their shape changes with symmetry. The acene series has only one narrow band which is broader in the phenes and becomes split into two or more bands in the strongly asymmetric hydrocarbons.

References

1. E. Clar, *Ber. Deut. Ges.*, **69**, 607 (1936); *Chem. Ber.*, **82**, 495 (1949);
   E. Clar, *Polycyclic Hydrocarbons*, *I*, Academic Press, New York,
   1964, p. 47.
2. E. Clar, *Polycyclic Hydrocarbons*, *I*, Academic Press, New York,
   1964, p. 50.
3. J. R. Platt, *J. Chem. Phys.*, **17**, 484 (1949); H. B. Klevens and
   J. R. Platt, *J. Chem. Phys.*, **17**, 470 (1949); W. Moffit, *J. Chem.
   Phys.*, **22**, 320 (1954).
4. R. M. Hochstrasser, *J. Chem. Phys.*, **33**, 459 (1960); R. Williams,
   J. Chem. Phys., **26**, 1186 (1957); H. Zimmermann and N. Joop,
   Z. Elektrochem., **64**, 1215 (1960); **65**, 61, 66, 138 (1961); R. H. Cox,
   H. W. Terry, Jr. and L. Harrison, *Tetrahedron Letters*, **50**, 4815
   (1971).

CHAPTER 5

Isomeric Hydrocarbons with a Different Number of Sextets

Having assessed the annellation effects in the acene and phene series and obtained a classification of absorption bands one can begin to compare hydrocarbons with different numbers of sextets and ascertain their influence on the electronic spectra. Just as branched and straight chains of C atoms cause different properties in aliphatic hydrocarbons the same is true for straight and branched annellated aromatic hydrocarbons, only in a more drastic way.

The general principle is that the stability of the isomers increases with the number of sextets. This also produces big shifts of the absorption bands to the violet. Heptacene (I) is green-black and cannot be obtained in a pure state because of its high reactivity. If one ring is arranged angularly the blue-green benzo-hexacene (II) is formed. It is already stable enough for the first band to be measured. This shifts by a very large amount (−1280 Å) if a second ring is brought into the angular position. Dibenzo-pentacene (III) has three sextets and is red. Passing to dibenzo-pentacene (IV) little change is observed because the number of sextets remains the same. Another big shift towards the violet is recorded in going to tribenzotetracene (V) which has four sextets, is yellow and considerably less reactive than tetracene itself. Arranging all the rings in such a way that five sextets are formed, and no double bond is left outside the sextets, gives the colourless tetrabenzanthracene (VI).[1] This is an anthracene in name only because it does not react with maleic anhydride to form an endo-cyclic adduct and is difficult to oxidize to a quinone.[2] This hydrocarbon belongs to the class of "fully benzenoid" hydrocarbons which have an extreme stability, it does not dissolve in concentrated sulphuric acid and shows an intense phosphorescence of long life at low temperature in solid solution.[3]

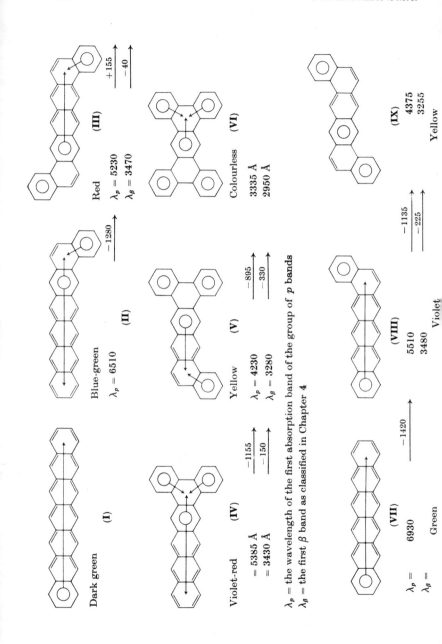

(I) Dark green

(II) Blue-green $\lambda_p = 6510$

(III) Red $\lambda_p = 5230$ $\lambda_\beta = 3470$ +155 −40

$\xrightarrow{-1280}$

(IV) Violet-red $= 5385$ Å $= 3430$ Å $\xrightarrow{-1155}$ $\xrightarrow{-150}$

(V) Yellow $\lambda_p = 4230$ $\lambda_\beta = 3280$ $\xrightarrow{-895}$ $\xrightarrow{-330}$

(VI) Colourless 3335 Å 2950 Å

λ_p = the wavelength of the first absorption band of the group of p bands
λ_β = the first β band as classified in Chapter 4

(VII) $\lambda_p = 6930$ $\lambda_\beta =$ Green $\xrightarrow{-1420}$

(VIII) 5510 3480 Violet $\xrightarrow{-1135}$ $\xrightarrow{-225}$

(IX) 4375 3255 Yellow

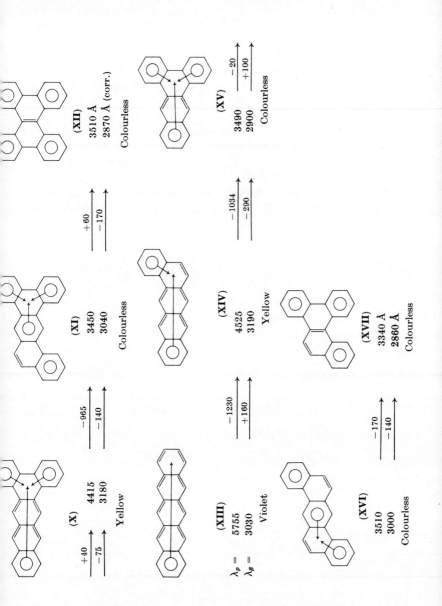

(XII)
3510 Å
2870 Å (corr.)
Colourless

(XV)
3490
2900
-20
$+100$
Colourless

(XI)
3450
3040
$+60$
-170
Colourless

(XIV)
4525
3190
-1034
-290
Yellow

(XVII)
3340 Å
2860 Å
Colourless

(X)
4415
3180
-965
-140
Yellow

(XIII)
$\lambda_p =$ 5755
$\lambda_\beta =$ 3030
-1230
$+160$
Violet

(XVI)
3510
3000
-170
-140
Colourless

The above sequence of events is repeated in a series beginning with the green hexacene (**VII**). Again a very strong shift towards shorter wavelength is recorded in going to the isomeric violet benzopentacene (**VIII**) which has two sextets. Passing to the yellow dibenzotetracene (**IX**) with three sextets brings a similar big shift. Dibenzotetracene (**X**) is little different because it also has three sextets. A large shift is again recorded in passing to the colourless tribenzanthracene (**XI**). There is one true double bond left outside the sextets which can be oxidized to an *o*-quinone. The isomeric dibenzochrysene (**XII**) also contains four sextets which forces the double bond into the position between the rings. This causes the oxidation to cleave the double bond with the formation of a cyclic ten-membered diketone. The principle can be generally applied. The violet pentacene (**XIII**) goes over into the yellow benzotetracene (**XIV**) with two sextets, accompanied by a very big shift towards shorter wavelength. A similar shift is recorded when passing to dibenzanthracene (**XV**). This has three sextets like the isomeric dibenzanthracene (**XVI**), both hydrocarbons being little different. A minor change is also observed in going to benzochrysene (**XVII**). The outer true double bond can be oxidized forming an *o*-quinone.

(**XVIII**)

$\lambda_p =$ 4710 $\xrightarrow{-1150}$

$\lambda_\beta = \begin{Bmatrix} 2930 \\ 2740 \end{Bmatrix} \xrightarrow{+25}$

Orange

(**XIX**)

3560 $\xrightarrow{-370}$

2860 $\xrightarrow{-190}$

Colourless

(**XX**)

3190 $\xrightarrow{-350}$

2670 $\xrightarrow{-100}$

Colourless

(**XXI**)

2840

2570

Colourless

Another series begins with the yellow tetracene (**XVIII**). The transition to tetraphene (benzanthracene) (**XIX**) brings the expected large shift towards shorter wavelength. Much smaller shifts are observed in going to the isomeric chrysene (**XX**) which also has two sextets. The most stable member of this series is the fully benzenoid triphenylene (**XXI**). It is neither sulphonated nor protonated by concentrated sulphuric acid.[4]

The above three series show that isomeric hydrocarbons with the same number of sextets give only minor spectral differences. They could be called sextet isomers. Nevertheless these differences are significant because it is always the hydrocarbon with three branches which has the β band more shifted towards shorter wavelength (see Chapters 7 and 8).

If the sextet is the symbol of benzenoid stability, then hydrocarbons which are built up entirely from rings with sextets must be the most stable amongst their isomers with the same number of rings. As a first approximation they can be considered condensed polyphenyls.

References

1. E. Clar and A. McCallum, *Tetrahedron*, **10**, 171 (1960).
2. P. Lambert and R. H. Martin, *Bull. Soc. Chim. Belg.*, **61**, 124 (1952).
3. E. Clar, *Polycyclic Hydrocarbons*, *I*, Academic Press, New York, 1964, p. 38.
4. E. Clar, *Polycyclic Hydrocarbons*, *I*, Academic Press, New York, 1964, p. 37.

CHAPTER 6

Fully Benzenoid Hydrocarbons

The great stability of benzene is also found in biphenyl (**I**). Triphenylene (**II**) is the most unreactive hydrocarbon among the isomers obtained by distillation of coal tar. If the triphenylene fraction is treated with concentrated sulphuric acid all the isomers are sulphonated except triphenylene. Dibenzopyrene (**III**) is a colourless hydrocarbon and does not show the reactivity of pyrene. Tribenzoperylene (**IV**) is also colourless and does not give a benzenogenic diene synthesis with maleic anhydride like perylene. Tetrabenzanthanthrene (**V**) and hexabenzocoronene (**VI**) are pale yellow hydrocarbons of great stability. In the mass spectrometer hexabenzocoronene forms a three-fold negative ion without fragmentation. Its melting point could not be determined because the melting-point tube melted long before the hydrocarbon.[1]

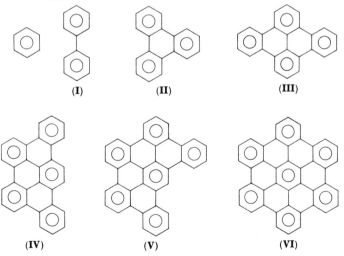

(I)   (II)   (III)

(IV)   (V)   (VI)

	(II)		(III)		(IV)	
$\lambda_p =$	2840	$\xrightarrow{440}$	3280	$\xrightarrow{460}$	3740	$\xrightarrow{300}$
$\lambda_\beta =$	2570	$\xrightarrow{310}$	2880	$\xrightarrow{120}$	3000	$\xrightarrow{240}$

	(VII)		(VIII)	
$\lambda_p =$	4040	$\xrightarrow{310}$	4350	Å
$\lambda_\beta =$	3240	$\xrightarrow{70}$	3310	Å

By comparison with the acene or phene series the shifts in the above series of triphenylene (**II**), dibenzopyrene (**III**), tribenzoperylene (**IV**), tetrabenzanthanthrene (**VII**) and tetrabenzoterrylene (**VIII**) are small. Calculated per π electron they have the smallest shifts found for all possible combinations of rings.

Another series starting with triphenylene (**II**), going over to tetrabenzanthracene (**IX**), tetrabenzopentacene (**X**) and tetrabenzoheptacene (**XI**) shows similar shifts, per π electron. It is remarkable that the above benzologues of pentacene (**X**) and heptacene (**XI**) do not show the high reactivity of the acenes but are extremely stable hydrocarbons. The only reason for this can be the presence of extra sextets.

If graphite were built up in this way, the black colour and the electric conductivity would appear very late.

The VB theory assumes nine Kekulé structures for triphenylene which are supposed to contribute equally to the ground state. Eight of these structures are contained in the formula with three sextets (**II**). The ninth structure **XII** has only one sextet; this is in the centre of the molecule and is written with three double bonds. The difference in aromatic energy between tetracene which has one sextet and the isomeric triphenylene amounts to 6–7 kcal. It is therefore difficult to see how structure **XII** could contribute equally with the others to form the ground state of triphenylene.

(II)   **(IX)**   **(X)**

$\lambda_p = 2840 \xrightarrow{495} 3335 \xrightarrow{105} 3440 \xrightarrow{}$

$\lambda_\beta = 2570 \xrightarrow{380} 2950 \xrightarrow{150} 3100 \xrightarrow{}$

(XI)

$\xrightarrow{260} 3700 \quad \text{Å}$

$\xrightarrow{190} 3290 \quad \text{Å}$

It is an advantage of the sextet formula **II** that it excludes this possibility.

If the fully benzenoid hydrocarbons are considered to be condensed polyphenyls this is not the whole truth. From the beginning it was assumed that two π electrons on the highest level can migrate from one ring to the other. Applied to triphenylene (**II**) this means that the central ring cannot be empty but must contain π electrons transferred from the neighbouring rings, as indicated by the arrows. Thus an induced sextet is formed which cannot be marked by a circle because this would give the wrong number of π electrons. The induced sextets provide an additional element of stability to these hydrocarbons which may surpass that of benzene.

This is supported by magnetic measurements. Triphenylene has a magnetic susceptibility which is greater than the one calculated

(II)   **(XII)**

from the three external benzene rings. This calculation is made by using the Pascal increments. Three times the susceptibility of benzene minus the atomic increment from six H atoms gives a molecular susceptibility of -149.2×10^{-6}. The measured susceptibility is -156.6×10^{-1}. The excess of -7.4×1^{-1} corresponds to the value for 0.4 rings.[2] This justifies the formula with the arrows (**II**).

In accordance with the above value it can be shown that the aromatic conjugation in triphenylene is not greater than in phenanthrene. The n.m.r. signals migrate with progressive annellation to lower fields. Thus the protons of benzene absorb at $2.73\,\tau$, the β protons of phenanthrene at 2.43 and $2.49\,\tau$, respectively, and the β protons of triphenylene at $2.42\,\tau$. One must conclude that the π electrons in triphenylene do not migrate beyond the central ring.

The fully benzenoid hydrocarbons have $6n$ π electrons, n being an integer. It becomes obvious that the most stable hydrocarbons do not follow Hückel's rule: $(2 + 4n)$ π electrons.[3] Therefore the latter must be strictly limited to monocyclic systems.

References

1. E. Clar, and C. T. Ironside, *Proc. Chem. Soc.* **1958**, 150; E. Clar, C. T. Ironside and M. Zander, *J. Chem. Soc.*, **1959**, 142.
2. H. Akamatu and M. Kinoshita, *J. Chem. Soc. Japan*, **32**, 774 (1959).
3. E. Hückel, *Grundzüge der Theorie ungesättigter and aromatischer Verbindungen*, Verlag Chemie, Berlin, 1938, p. 72, 75.

CHAPTER 7

Asymmetric Annellation Effect in the Dibenzacene Series

Passing from naphthalene (**I**) to phenanthrene (**II**) one finds a normal annellation effect amounting to +300 Å and indicating full aromatic conjugation. The two sextets in phenanthrene enforce the fixation of the double bond in the 9.10-position. If another ring with an inherent sextet is annellated in going to triphenylene (**III**) the effect is very small (+60 Å). It means that formally symmetric annellation produces asymmetric annellation effects. The reason for this must be in the central ring with the induced sextet. This is confirmed by the next series beginning with anthracene (**IV**). This has still only one sextet. One angular ring with one sextet brings a normal annellation effect of +355 Å and a fixation of the double bond in tetraphene (benzanthracene) (**V**). Fusing a ring to the fixed double bond results in a very small shift to shorter wavelength of −10 Å when dibenzotetracene (**VI**) is formed. This asymmetric annellation effect is also found in the series starting from tetracene (**VII**). The first ring again produces a normal shift of +420 Å. This causes the fixation of the double bond in benzotetracene (**VIII**). The next ring in going to dibenzo-tetracene (**IX**) brings a very small shift of −30 Å. The same annellation effects are observed in the pentacene series **X**, **XI** and **XII**, the corresponding shifts being +450 and −50 Å, respectively.

This asymmetric annellation effect is always observed if three linear branches of rings are fused to a central ring. Then only the two longest branches are in aromatic conjugation, at least during the time of light absorption. No exception to this rule has been found so far. It is based on the fixation of a double bond which cuts off the conjugation of any branch attached to it, just as is the case with crossed conjugated systems of olefinic hydrocarbons.[1]

(I)

(II)

(III)

$\lambda_\beta = 2210$ $\xrightarrow{+300}$ 2510 $\xrightarrow{+60}$ 2570 Å

$\sqrt{\lambda_\beta} = 47.01$ $\xrightarrow{+3.09}$ 50.10 $\xrightarrow{+0.60}$ 50.70 $\sqrt{}$ Å

(IV)

(V)

(VI)

$\lambda_\beta = 2515$ $\xrightarrow{+355}$ 2870 $\xrightarrow{-10}$ 2860 Å

$\sqrt{\lambda_\beta} = 50.15$ $\xrightarrow{+3.43}$ 53.58 $\xrightarrow{-0.10}$ 53.48 $\sqrt{}$ Å

(VII)

(VIII)

(IX)

$\lambda_\beta = 2740$ $\xrightarrow{+420}$ 3160 $\xrightarrow{-30}$ 3130 Å

$\sqrt{\lambda_\beta} = 52.35$ $\xrightarrow{+3.87}$ 56.22 $\xrightarrow{-0.27}$ 55.95 $\sqrt{}$ Å

(X)

(XI)

$\lambda_\beta = 3030$ $\xrightarrow{+450}$ 3480

$\sqrt{\lambda_\beta} = 55.05$ $\xrightarrow{+3.95}$ 50.00

(XII)

$\xrightarrow{-50}$ 3430 Å

$\xrightarrow{-0.42}$ 58.58 $\sqrt{}$ Å

The n.m.r. spectra support the assumption of fixed double bonds in the phenes. The coupling constant found for the double bond is about $J = 9$ Hz. This is more than for other aromatic bonds. A rather superficial comparison with classically fixed double bonds is possible in 1,2-dihydronaphthalene (**XIII**)[2] and perinaphthene (phenalene) (**XIV**)[3] where the corresponding coupling constants for the double bonds are 9.59 and 9.6 Hz, respectively.

(**XIII**)   (**XIV**)

There is another possibility of testing the assumed double-bond fixation. The equal protons of a CH_3 group couple with a neighbouring aromatic proton if both are connected through a double bond. This causes the CH_3 protons to form a doublet whilst the absence of a double bond prevents coupling and the CH_3 protons then form a sharp singlet. The separation of the CH_3 doublet is usually 1 Hz as shown by the following examples:

1 Hz   1 Hz   1 Hz

1 Hz

In methylchrysene (**XV**) the CH_3 doublet shows the normal separation of 1 Hz. One must assume that the ring with the CH_3 has the character of the middle ring of phenanthrene. This is not so in dimethylchrysene where the movement of the sextets causes the CH_3 groups to be attached alternatively to a double bond (**XVIa**) and to a sextet (**XVIb**). This reduces the separation to 0.7 Hz. One can easily see that the method can be applied to very intricate fine-structure problems.

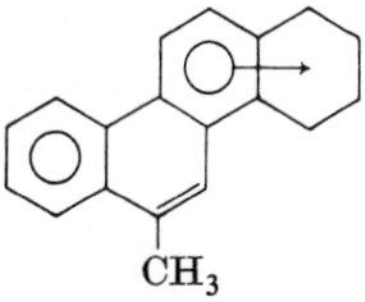

(XV) 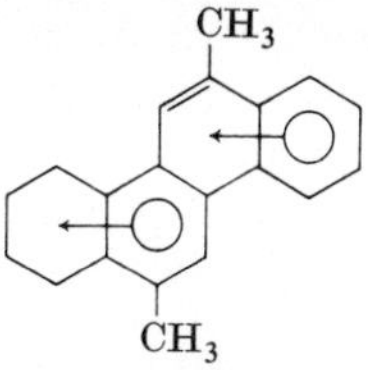　　　　(XVIa) 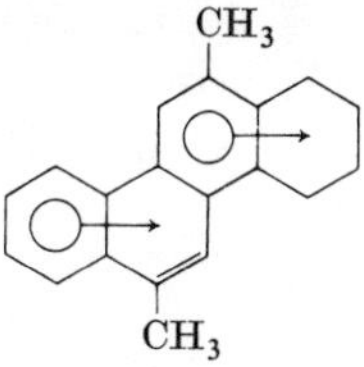　　　　(XVIb)

If there is no double-bond character between the CH₃ group and the adjacent aromatic proton no splitting of the CH₃ signal can be observed, e.g. in 1,5-dibromo-2,6-dimethylnaphthalene (**XVII**).[4]

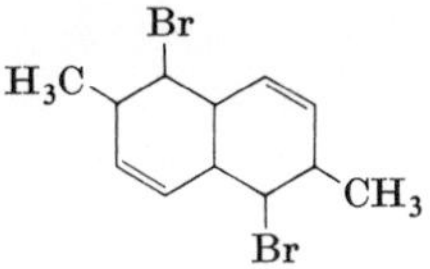

(**XVII**)

There is an interesting possibility of observing asymmetric annellation effects through the chemical shifts of the sterically hindered protons (bay protons). If there is any annellation effect, the additional ring current should produce a shift to lower field (i.e. higher frequency). These shifts are measured in Hz ex TMS (tetramethylsilane) at 100 MHz in CS_2 solution.

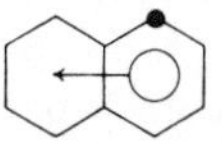

(**I**) 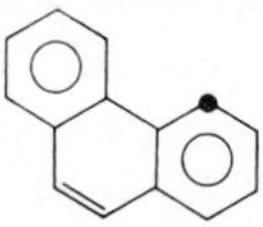　　　　(**II**) 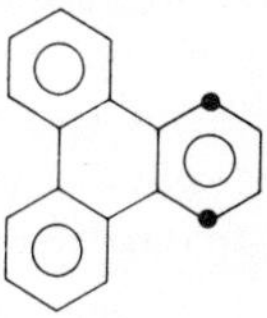　　　　(**III**)

$$777 \xrightarrow{+80} 850 \xrightarrow{+3} 853 \text{ Hz ex TMS}$$

There is a big shift of the marked protons to lower field in going from naphthalene (**I**) to phenanthrene (**II**) because these protons come under the influence of the ring current of the additional ring. A further ring in triphenylene (**III**) produces only a very small shift of +3 Hz indicating that the three external rings are hardly in aromatic conjugation. This was also found in the electronic spectra.

A similar observation can be made in passing from anthracene
(**IV**) to tetraphene (benzanthracene **V**) and dibenzanthracene (**VI**).
There is again first a big positive shift and then a small negative
shift for the protons marked with points and crosses.

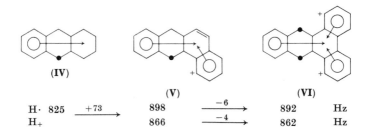

H·	825	+73 →	898	−6 →	892	Hz
H+			866	−4 →	862	Hz

In the tetracene series the same observation can be recorded.

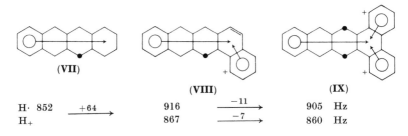

H·	852	+64 →	916	−11 →	905	Hz
H+			867	−7 →	860	Hz

The trend of the experimental values is exactly the same as in
the electronic spectra (p. 38) the only difference being that during
the very short time of light absorption the molecules are found
in an asymmetric electronic state, whilst in the n.m.r. spectra
formally magnetic equivalent protons must give averaged values.
This is also shown by the following series:

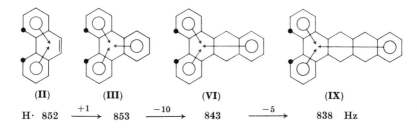

	(II)		(III)		(VI)		(IX)	
H·	852	+1 →	853	−10 →	843	−5 →	838	Hz

Between phenanthrene (**II**) and triphenylene (**III**) there is again a very small positive shift which is followed by a negative shift, i.e. to higher field. One comes to the conclusion that increasing the length of the branch does not contribute to the ring current in the central ring, but rather withdraws electrons from it.[5] It is certainly not possible for all three branches to be in aromatic conjugation at the same time.

References

1. E. Clar, *Tetrahedron*, **5**, 98 (1959); **6**, 335 (1959); **9**, 202 (1960).
2. M. A. Cooper, D. D. Ellemann, C. D. Pearce and S. L. Manott, *J. Chem. Phys.*, **53**, 2343 (1970); M. J. Cook, A. R. Katritzky, F. C. Pemington and B. M. Semple, *J. Chem. Soc. (B)*, **1969**, 523.
3. M. Prinzbach, V. Freudenberger and U. Schneidegger, *Helv.Chim. Acta*, **50**, 1087 (1967).
4. E. Clar, B. A. McAndrew and Ü. Sanigök, *Tetrahedron*, **26**, 2099 (1970).
5. E. Clar and C. C. Mackay, Future communication.

CHAPTER 8

The Starphene Series

The starphenes are the higher members in the series of linearly annellated triphenylenes. Here again a strong element of asymmetry can be observed. There is a normal shift in passing from triphenylene (**I**) to dibenzanthracene (**II**) as in the acene and phene series. A similar shift is recorded in going to hexastarphene (1.2.2) (**III**). However, another benzene ring fused to the third branch causes a negative shift of -20 Å (**IV**).[1]

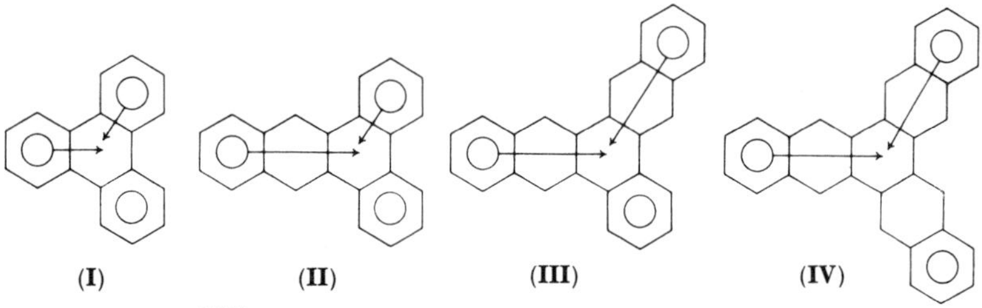

| (I) | (II) | (III) | (IV) |

$\lambda_\beta = 2570 \quad \xrightarrow{+300} \quad 2870 \quad \xrightarrow{+270} \quad 3140 \quad \xrightarrow{-20} \quad 3210$ Å

There cannot be an aromatic conjugation between all three branches during the time of light absorption. This is marked by writing only two arrows into the system. The two longest branches determine the long wavelength part of the spectrum. There may also be combinations between two shorter branches; these would absorb at shorter wavelengths.

The asymmetry can be presented in a more drastic way by starting with benzene and fusing two rings to it. This causes a shift of $+663$ Å in going to anthracene. Passing to pentaphene (**V**) another similar shift of $+630$ Å is produced. However, the third branch causes a shift of -25 Å towards shorter wavelengths.

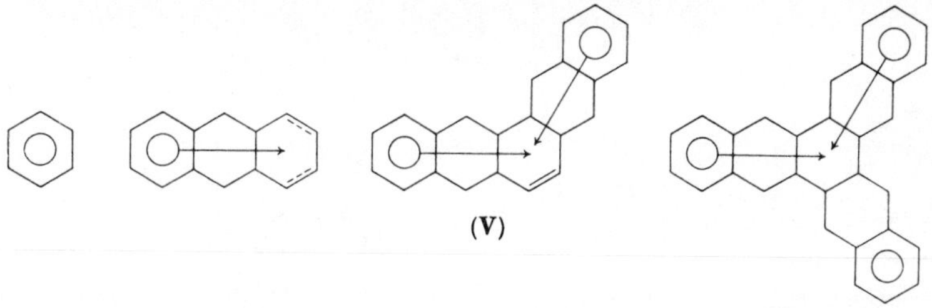

$$\lambda_\beta = 1852 \xrightarrow{+663} 2513 \xrightarrow{+630} 3145 \xrightarrow{-25} 3120 \text{ Å}$$

(V)

(IV)

$$\xrightarrow{+70}$$

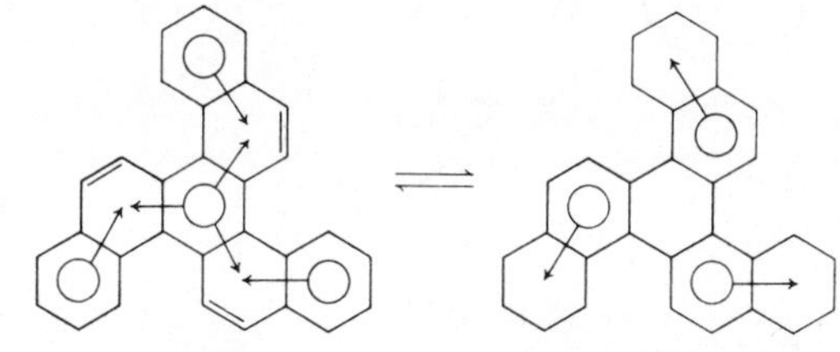

(VIa)   (VIb)

3190 Å

If three rings are angularly annellated to triphenylene the hydrocarbon **VI**[2] is obtained which does not belong to the star-phene type. In one form **VIa** the central ring can achieve an inherent sextet which in conjunction with the form **VIb** can promote aromatic conjugation between all three branches. The result is a shift of 70 Å by comparison with heptastarphene (trinaphthylene, **IV**).

A particularly striking example of asymmetric annellation is presented in Figure 11. Starting with heptaphene proceeding to octastarphene(1.3.3), nonastarphene(2.3.3) and decastarphene (3.3.3) no shifts towards the red can be observed; there is a rather small shift to shorter wavelength. A three-fold conjugation of three branches should produce a red shift of more than 1000 Å, which is not the case.[3]

The n.m.r. spectra in the starphene series support the view that the third branch cannot be in aromatic conjugation with the other two. The protons marked H · shift to higher field with progressive annellation of the third branch onto pentaphene (**V**). The opposite effect should be expected if the third branch were to increase the ring current through aromatic conjugation. The protons marked

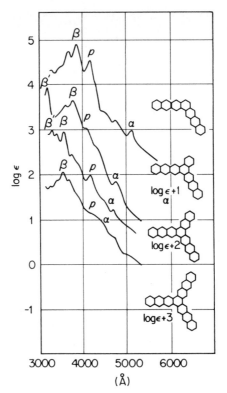

Figure 11. Absorption spectra of heptaphene, octastarphene (1.3.3), nonastarphene(2.3.3) and decastarphene(3.3.3)

H+ shift to lower field until the values for H· and H+ are the same, both being at higher field than the H· in pentaphene (**V**):

$$
\begin{array}{cccccc}
\text{H·} & 907 & \xrightarrow{-5} & 902 & \xrightarrow{-7} & 895 \quad \text{Hz ex TMS} \\
\text{H+} & 810 & \xrightarrow{+72} & 882 & \xrightarrow{+13} & 895 \quad \text{Hz ex TMS}
\end{array}
$$

(**V**)   (**III**)   (**IV**)

The value of 895 Hz for all the meso protons is an averaged value which, different from light absorptions, is caused by the relatively long time of an n.m.r. record.

In contrast to the starphene series aromatic conjugation does take place, although on a reduced scale, if benzene rings are annellated in a zig-zag way. The p and β bands are recorded below for phenanthrene (**VII**), chrysene (**VIII**), picene (**IX**) and fulminene (**X**):

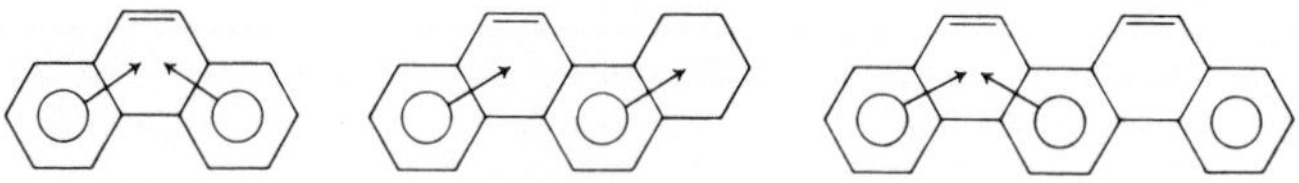

(**VII**)   (**VIII**)   (**IX**)

$\lambda_p = 2925 \xrightarrow{+265} 3190 \xrightarrow{+100} 3290$

$\lambda_\beta = 2510 \xrightarrow{+160} 2670 \xrightarrow{+200} 2870$

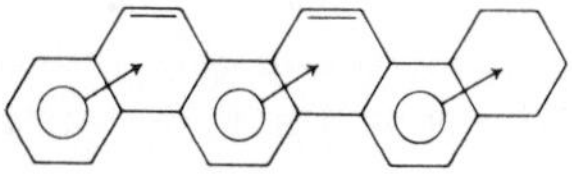

(**X**)

$\xrightarrow{+110}$ 3400 Å

$\xrightarrow{+80}$ 3950 Å

Conjugation takes place as indicated by the arrows, showing that the π-electron transport is more complicated and less efficient than in the acene or phene series.[4]

References

1. E. Clar, A. McCallum and R. A. Robertson, *Tetrahedron*, **18**, 1471 (1962).
2. W. H. Laarhoven and J. A. M. van Broekhoven, *Tetrahedron Letters*, **1**, 73 (1970).
3. E. Clar and A. Mullen, *Tetrahedron*, **24**, 6719 (1968).
4. E. Clar and O. Kühn, *Liebigs Ann.*, **601**, 181 (1956).

Asymmetric Annellation in the Tetrabenzacene Series

Just as the formula with the four valences of carbon directed to the corners of a tetrahedron points to the existence of stereoisomers, writing the formula of tetrabenzanthracene (**III**) with the sextets and arrows predicts asymmetric annellation effects in passing from benzene (**I**) to triphenylene (**II**) and tetrabenzanthracene (**III**). This is indeed the case, the first shift being +718 and the second +355 Å.

Applying the more accurate scale of $\sqrt{\text{Å}}$ as in the acene and phene series one can subtract the effect of the annellation of the second biphenyl complex from triphenylene which must leave an empty ring "E". This subtraction from the value for triphenylene yields a β band at 2237 Å which is very near the β band found for naphthalene which is at 2210 Å. This shows that in the course of the annellation process the aromatic system of benzene was extended to a naphthalene system but not to an anthracene system. This is exactly described by the arrows in **III**.

The annellation of a biphenyl complex to naphthalene (**IV**) brings a big shift in passing to 1.2.3.4-dibenzanthracene (**V**) and a smaller shift in going to tetrabenzotetracene (**VI**). The application of the above subtraction rule gives a β band at 2491 Å which is almost identical with the β band of anthracene (2515 Å in alcoholic solution). It is obvious that the first annellation extended the aromatic conjugation of naphthalene (**IV**) to an anthracene system in **V** and that the second fusion of a biphenyl system to **V** produced an empty ring "E" in **VI**.[1]

In the next highest series, anthracene (**VII**), dibenzotetracene (**VIII**) and tetrabenzopentacene (**IX**) an analogous observation can be made. The subtraction rule yields the β band of tetracene. This

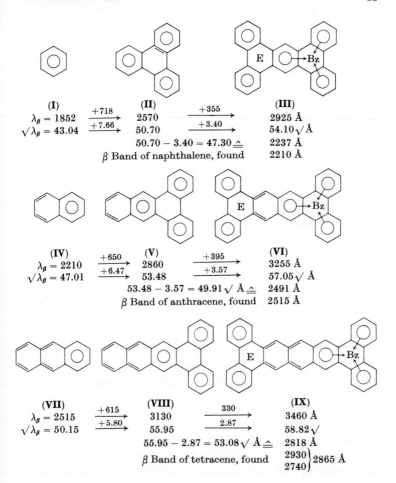

(I)
$\lambda_\beta = 1852$
$\sqrt{\lambda_\beta} = 43.04$

$\xrightarrow[+7.66]{+718}$

(II)
2570
50.70

$\xrightarrow[+3.40]{+355}$

$50.70 - 3.40 = 47.30 \triangleq$
β Band of naphthalene, found

(III)
2925 Å
$54.10 \sqrt{}$ Å
2237 Å
2210 Å

(IV)
$\lambda_\beta = 2210$
$\sqrt{\lambda_\beta} = 47.01$

$\xrightarrow[+6.47]{+650}$

(V)
2860
53.48

$\xrightarrow[+3.57]{+395}$

$53.48 - 3.57 = 49.91 \sqrt{}$ Å $\triangleq$
β Band of anthracene, found

(VI)
3255 Å
$57.05 \sqrt{}$ Å
2491 Å
2515 Å

(VII)
$\lambda_\beta = 2515$
$\sqrt{\lambda_\beta} = 50.15$

$\xrightarrow[+5.80]{+615}$

(VIII)
3130
55.95

$\xrightarrow[2.87]{330}$

$55.95 - 2.87 = 53.08 \sqrt{}$ Å $\triangleq$
β Band of tetracene, found

(IX)
3460 Å
$58.82 \sqrt{}$
2818 Å
$\left.\begin{array}{c}2930\\2740\end{array}\right\}2865$ Å

comparison indicates that the anthracene system is extended to a tetracene system and not to an aromatic pentacene system in **IX**, thus producing an induced benzenoid ring (Bz) and an empty ring "E".

It was shown in Chapters 7 and 8 that the third branch in triphenylene and its linear benzologues is not in aromatic conjugation with the two other branches. Therefore it should not influence the results of the above asymmetric annellations if the third branch were to be removed. In other words the effect should be the same if instead of biphenyl systems styrene systems were annellated. This is found to be the case:

$$\text{(I)}$$
$$\lambda_\beta = 1852 \quad \xrightarrow{+658} \quad 2510 \quad \xrightarrow{+460} \quad 2970 \text{ Å}$$
$$\sqrt{\lambda_\beta} = 43.04 \quad \xrightarrow{+7.06} \quad 50.10 \quad \xrightarrow{+4.40} \quad 54.50\sqrt{} \text{ Å}$$
$$50.10 - 4.40 = 45.70 \qquad \triangleq \qquad 2089 \text{ Å}$$

β Band of naphthalene, found   2210 Å

$$\text{(XI)} \qquad\qquad\qquad\qquad \text{(XII)}$$
$$\lambda_\beta = 1852 \quad \xrightarrow{+658} \quad 2510 \quad \xrightarrow{+335} \quad 2845 \text{ Å}$$
$$\sqrt{\lambda_\beta} = 43.04 \quad \xrightarrow{+7.06} \quad 50.10 \quad \xrightarrow{+3.25} \quad 53.35\sqrt{} \text{ Å}$$
$$50.10 - 3.25 = 46.85\sqrt{} \text{ Å} \quad \triangleq \qquad 2195 \text{ Å}$$

β Band of naphthalene, found   2210 Å

The fusion of a styrene complex to benzene (**I**) leads to phenanthrene with a shift of +658 Å and a second styrene gives 1.2,5.6-dibenzanthracene (**X**) with a shift of +460 Å. The shift difference rule shows that benzene has been extended to a naphthalene system in **X** and not to an anthracene system.

The process can be carried out in a different way leading to phenanthrene (**XI**) and then to picene (**XII**). Here the extension also yields a naphthalene system with a similar result for the above calculation. The slight differences can be explained by the differences in the electronic fine structure in the central ring of dibenzanthracene (**X**) and picene (**XII**).

The dibenzotetracenes, **XIII** and **XIV**, can be obtained by asymmetric annellation from naphthalene and anthracene, respectively. The shift difference rule gives the correct answer in all three cases.

The rule can be applied to the extreme case starting from ethylene. For this process it does not matter whether the absorption

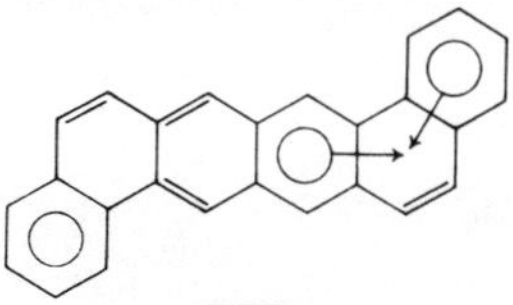

(XIII)
3255 Å
57.05 √ Å

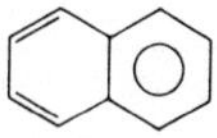

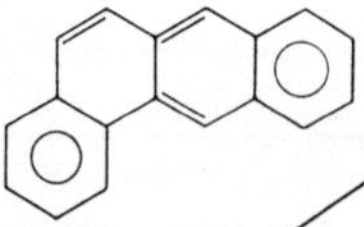

$\lambda_\beta = 2210$ $\xrightarrow{+690}$ 2900
$\sqrt{\lambda_\beta} = 47.01$ $\xrightarrow{+6.85}$ 53.86

+355
+3.19

+360
+3.24

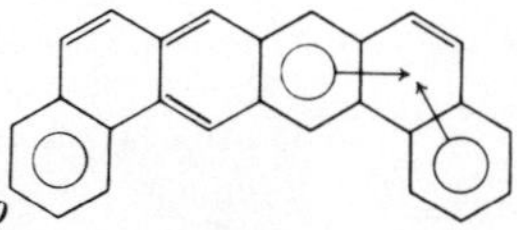

(XIV)
3260 Å
57.10 √ Å

$53.86 - 3.19 = 50.67 \sqrt{} \text{ Å} \triangleq 2567 \text{ Å}$
$53.86 - 3.24 = 50.62 \sqrt{} \text{ Å} \triangleq 2562 \text{ Å}$
β Band of anthracene, found   2534 Å (corrected for benzene solution)

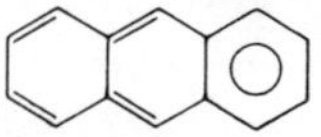

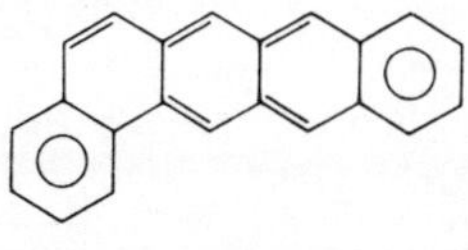

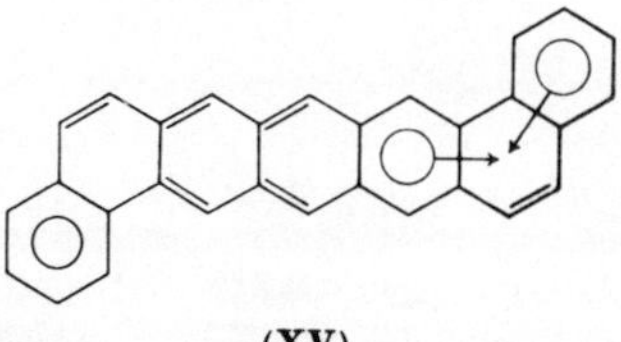

(XV)
3505 Å
59.20 √ Å

$\lambda_\beta = 2515$ $\xrightarrow{+675}$ 3190 $\xrightarrow{+315}$
$\sqrt{\lambda_\beta} = 50.10$ $\xrightarrow{+6.34}$ 56.49 $\xrightarrow{+2.71}$

$56.49 - 2.71 = 53.78 \sqrt{} \text{ Å} \triangleq 2892 \text{ Å}$

β Band of tetracene, found $\begin{Bmatrix} 2930 \\ 2740 \end{Bmatrix} \triangleq 2835 \text{ Å}$ (corrected for benzene solution)

Alcoholic solution

of ethylene can be compared with naphthalene and chrysene (**XVI**). Only the difference between the two latter hydrocarbons is relevant. The first fusion of a styrene complex leads to naphthalene and the second to chrysene (**XVI**). The subtraction or shift difference rule can in this case be applied to the α and β bands. It gives the correct figures for the α and β bands of benzene in very good agreement with the experimental values.[1]

(XVI)

$$\lambda_\beta = 1976 \xrightarrow{+234} 2210 \xrightarrow{+460} 2670 \text{ Å}$$
$$\sqrt{\lambda_\beta} = 44.45 \xrightarrow{+2.56} 47.01 \xrightarrow{+4.67} 51.68 \sqrt{}\text{ Å}$$

$$\lambda_{a^2}\ 3110 \xrightarrow{+490} 3600 \text{ Å}$$
$$\sqrt{\lambda_{a^2}}\ 55.78 \xrightarrow{+4.22} 60.00 \sqrt{}\text{Å}$$

$55.78 - 4.22 = 51.56 \sqrt{}\text{ Å} \triangleq 2658 \text{ Å}$

α Band of benzene, found   2604 Å (in alcohol)

$47.01 - 4.67 = 42.34 \sqrt{}\text{ Å} \triangleq 1792 \text{ Å}$

β Band of benzene, found 1790 Å (vapour)

The n.m.r. spectra show also a marked asymmetric annellation effect:

	(I)		**(XI)**		**(X)**	
H·	727	$\xrightarrow{+53}$	780	$\xrightarrow{+128}$	908	Hz ex TMS
H₊	727	$\xrightarrow{+123}$	850	$\xrightarrow{+58}$	908	Hz ex TMS

The above series involves "bay" protons. However, an asymmetric effect is also observed with no overcrowded bay protons in the next higher series[2]:

	(IV)				**(XIII)**	
H·	770	$\xrightarrow{+49}$	819	$\xrightarrow{+32}$	851	Hz ex TMS
H₊	770	$\xrightarrow{+128}$	898	$\xrightarrow{+22}$	920	Hz ex TMS

The n.m.r spectra are of particular importance in this context because only the ground state of the molecule is involved in the measurement.

References

1. E. Clar, *Tetrahedron*, **5**, 98 (1959); **6**, 355 (1959); **9**, 202 (1960).
2. E. Clar, Future communication.

CHAPTER 10

The Subdivision of the Aromatic Sextet

In the preceding chapters numerous facts have been reported which point to the existence of a true double bond in any benzenoid ring (**I**). This must be on the lowest of three π electron levels. It is mobile in benzene, partly restricted in benzene derivatives and in naphthalene derivatives in any $\alpha-\beta$ position as shown in formula **II**. The two sextets in phenanthrene enforce a localization in the 9.10-position (**III**). The double bond in the central ring of triphenylene can be in any one of three positions, **IVa**, **IVb** or **IVc**. (The other double bonds are not marked.*) The double bond between two rings must always exclude one ring from aromatic conjugation. For this reason structure **IVd** appears unacceptable, since it has only one sextet, which is in the central ring.

Fixed double bonds of the phene type (**III**) have been proved by the coupling of the CH_3 protons with the neighbouring aromatic proton. Localization of a double bond between two rings is completely restricted to 1.2,7.8-dibenzochrysene (**V**) and some of its derivatives. This becomes obvious in the very exceptional oxidation of this central double bond to the diketone **VI**.

The existence of the two π electrons on the second level which are localized within the ring has been proved by the result of asymmetric annellation. If these electrons are not present in any benzenoid ring no induced sextet can be formed and the ring must remain empty "E".

There are also two π electrons on the third level which can migrate between the rings as shown in naphthalene (**II**). These can also fill an empty ring and form an induced sextet as in phenanthrene (**III**). This ring would otherwise remain empty.[1]

---

* All hexagons symbolize aromatic rings.

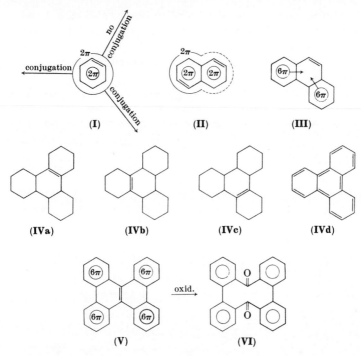

(I)   (II)   (III)

(IVa)   (IVb)   (IVc)   (IVd)

(V)   (VI)

With this model of π-electron distribution in mind one can try to explain some features in the n.m.r. spectra of CH_3 derivatives which otherwise defy interpretation. The CH_3 signals are particularly suitable for testing the double-bond character of adjacent bonds because they are well outside the other aromatic proton absorption in the n.m.r. spectra and are usually in the range between 2 and 3 p.p.m., 7–8 τ or 200–300 Hz from TMS at 100 HMz.

It is an amazing fact that the *para*-CH_3 signals in the first annellation series in Figure 12 do not show a splitting due to neighbouring double bonds. No splitting of the CH_3 signal can be observed in *p*-xylene, whilst 1,4-dimethylnaphthalene and 1,4-dimethylanthracene show doublets with a separation of 0.2 and 0.3 Hz, respectively. Decoupling of the aromatic protons, as marked with points, also shows that the degree of coupling is very small and is not consistent with 50 per cent double-bond character in the bond adjacent to the CH_3 group, as required by the Kekulé structures. This can be attributed to one fixed double bond as indicated by the formulae for dimethylnaphthalene and dimethyl-anthracene, or in the case of p-xylene in two equal positions as marked by dashed lines. The other delocalized π electrons are not symbolized in the formulae.

Quite different results are observed in the second annellation series in Figure 12. The CH_3 band of o-xylene appears as a singlet; decoupling, marked by points, has very little effect. The CH_3 signal in 2,3-dimethylnaphthalene and 2,3-dimethylanthracene forms a

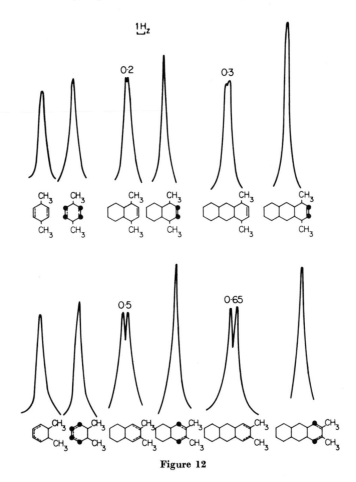

Figure 12

doublet with a separation of 0.5 and 0.65 Hz, respectively. Decoupling has a strong effect. Whilst the double bond in *o*-xylene can be in three non-adjacent bonds, only half a true double bond appears to be adjacent to the CH_3 groups in the former hydrocarbons, as marked by dashed lines.

The correctness of the assumption that only one true double bond is shared between the two α–β positions in 2,3-dimethylnaphthalene, as shown by the dashed line in Figure 12, second series, can be demonstrated by the CH_3 bands of 1-bromo-2,3-dimethylnaphthalene, Figure 13. The low field CH_3 signal (position 2) is a singlet, little sharpened by decoupling H_4. However, the signal of the high field CH_3 (position 3) is a doublet with a separation of 0.9 Hz, which is almost double the separation of the doublet in 2,3-dimethylnaphthalene (0.5 Hz). One must assume that the double bond is fixed as shown in Figure 13. The black area indicates the effect of decoupling H_4, which produces a singlet.

A shifting of the double bond can also be observed in 4-bromo-1-methylnaphthalene (Figure 13) by comparison with 1,4-dimethylnaphthalene (Figure 12). Whilst the latter shows no double bond adjacent to CH_3, the small splitting of 0.2 Hz being about the same as between CH_3 and a proton in the *meta* position, there is a very strong separation of 0.95 of the CH_3 doublet recorded in

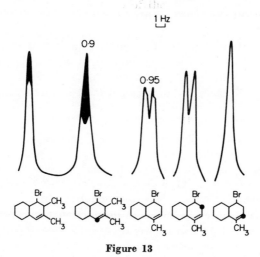

Figure 13

4-bromo-1-methylnaphthalene (Figure 13). There appears to be a double-bond fixation as shown in the formulae. Decoupling of the protons marked with points proves the origin of the coupling (0.95) and *meta* coupling (0.25 Hz). Removal of the *meta* coupling clears up the doublet and decoupling the *ortho* proton transforms it into a singlet.

In the first case in Figure 13 there is an additivity of parts of a true double bond recorded and in the second case a shifting of the double bond under the influence of bromine.[2] In many other cases bromine proves to be a very valuable agent for the repulsion of double bonds from the neighbourhood. The aldehyde group also has a strong influence in the same direction. 1-Methylnaphthalene (**VII**) has a complicated CH_3 signal which is the result of multiple coupling.[3] This can be drastically simplified by the introduction of an aldehyde group in 1-methylnaphthalene-4-aldehyde (**VIII**). This shows only a CH_3 doublet with a separation of 0.9 Hz. A similar concentration of double bond is found in 1-chloro-4-methylanthracene (**IX**) where the doublet separation is 1 Hz.

(**VII**)   (**VIII**)   (**IX**)

The additivity of parts of a double bond and the shifting of double bonds is completely incompatible with the Kekulé structures, because the repulsion of one double bond in a Kekulé structure would cause the approach of another and the assumption of an additivity of double-bond character under the influence of bromine would appear senseless.

The non-adjacent double bond of *o*-xylene (**X**) can be shifted under the influence of bromine in such a way that parts of a double bond become adjacent to the CH_3 and produce a doublet in 4,5-dibromo-*o*-xylene (**XI**) with a separation of 0.3 Hz (Figure 14). The CH_3 signal of durene (**XII**) is a doublet with a separation of 0.25–0.3 Hz (Figure 14). Since this results from two three-centred bonds (marked by dashed lines) the total degree of double-bond localization must be rather high.

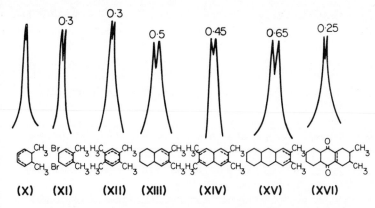

Figure 14. Comparison of CH_3 signal in the n.m.r. spectra of symmetric di- and tetramethyl derivatives

The doublet in 2,3-dimethylnaphthalene (**XIII**) (Figure 14) has a separation of 0.5 Hz. This is reduced to 0.45 Hz in 2,3,6,7-tetramethylnaphthalene (**XIV**). It is reasonable to assume that those circular delocalized π electrons which can migrate between the two rings are to some extent pushed into the non-substituted ring in 2,3-dimethylnaphthalene thus making the ring with the CH_3 groups less benzenoid and more reactive. This cannot be the case in 2,3,6,7-tetramethylnaphthalene (**XIV**) and one must conclude that these circular delocalized π electrons have a slight counteracting effect to the tendency to localize the true double bond adjacent to the β-methyl groups.

The bromination of tetramethylnaphthalene (**XIV**) gives 1,5-dibromo-2,3,6,7-tetramethylnaphthalene. The doublet in the n.m.r. spectrum shows an increased separation of 0.65 Hz, thus indicating a concentration of the double bonds in the 4,8-position.

The influence of the migrating electron pair on the highest level must be further reduced in 2,3-dimethylanthracene (**XV**) by comparison with 2,3-dimethylnaphthalene (**XIII**). In fact a separation of 0.65 Hz is found in the former case. In passing from benzene to naphthalene and anthracene derivatives a distinct annellation effect is thus observed.

An effect which is contrary to the effect of bromine can be observed in 2,3-dimethylanthraquinone (**XVI**). The electronic deficiency of the CO groups attracts the double bond partly away from the CH_3 groups as indicated by the dotted line in **XVI**. The CH_3 doublet separation of 0.65 Hz in the hydrocarbon is therefore reduced to 0.25 Hz in 2,3-dimethylanthraquinone (**XVI**).

There is a CH_3 signal with a separation of 0.4 Hz in 2,3-dimethyl-triphenylene (**XVII**). This value is lower than in 2,3-dimethyl-naphthalene (0.5 Hz) and in 2,3,6,7-tetramethylnaphthalene (0.45 Hz) and slightly higher than in dibromo-*o*-xylene and durene (0.3 Hz). Triphenylene is best formulated with three aromatic sextets in the external rings. There is a lower electronic density in the central ring. It is therefore understandable that the CH_3 splitting of the CH_3 doublet is between the above-mentioned benzene and naphthalene derivatives.

No splitting can be observed in the CH_3 signal of 8,9-dimethyl-fluoranthene (**XVIII**). Decoupling of the adjacent aromatic protons has very little effect as shown by the black area of the signal. One must therefore assume a similar electronic structure (Figure 14) for the ring attached to the CH_3 groups as in *o*-xylene (**X**). This appears reasonable, since 7,10-dimethylfluoranthene also shows no CH_3 doublet but a rather sharp singlet. In this case the ring attached to the CH_3 groups must be closely related to *p*-xylene.

6,7-Dimethyl-1.2,3.4-dibenzanthracene (**XIX**) (Figure 15) shows a CH_3 doublet with a separation of 0.55 Hz. This value is lower than

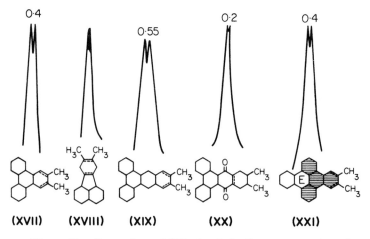

Figure 15. Comparison of CH_3 signals in the n.m.r. spectra of symmetric dimethyl derivatives

in 2,3-dimethylanthracene (**XV**). In passing to 6,7-dimethyl-1.2,3.4-dibenzanthraquinone (**XX**), a reduction of the CH_3 splitting from 0.55 Hz in the case of the hydrocarbon to 0.2 Hz in the quinone is recorded. This change is quite similar to that going from 2,3-dimethylanthracene to 2,3-dimethylanthraquinone and must also be related to the attraction of the double bond into the position between the electronically deficient CO groups of the quinone (dotted line in **XX**).

Dimethyldibenzopyrene (**XXI**) shows a CH_3 doublet with a separation of 0.4 Hz (**XXI**) (Figure 15) which is exactly the same value as in 2,3-dimethyltriphenylene (**XVII**) (Figure 15). This leads to the assumption that the aromatic conjugation of the triphenylene complex (marked with shadow) and an empty ring "E" dominate the electronic structure of **XXI**. If the two CH_3 groups were connected with the phenylene ring adjacent to the empty ring "E" then no splitting of the CH_3 signal should be observed as in *o*-xylene. In the case of the electronic structure resonating between these two structures the CH_3 splitting should be only about halfway between 0 and 0.4 Hz, which is obviously not the case. An asymmetric electronic structure for dibenzopyrene and its higher benzologues will be discussed on p. 72.

The CH_3 signal of toluene is a broad unresolved multiplet.[5] It is simplified in 4-bromotoluene (**XXII**) (Figure 16) to a septet produced by an overlapping triplet of triplet. That the double bond is located as a three-centred bond adjacent to CH_3 is shown by decoupling of the *meta*-protons (marked with points) (**XXIII**). Then the septet is simplified to a clear triplet with a separation of 0.6 Hz. In 2,4,5-tribromotoluene (**XXIV**) the CH_3 signal is a double doublet. Decoupling of the proton H_3 produces a CH_3 doublet with a separation of 0.95 Hz. This is the highest value observed in toluene derivatives and is not much different from the doublet in 9-methylphenanthrene (1.0 Hz).[4]

This shows that the electronic structure of the benzene ring is not principally different from any other benzenoid ring, only the mobility of the one true double bond is greater. If the double bond in 2,4,5-tribromotoluene is almost completely fixed then this cannot be the case to the same extent in 2,3,4,5-tetrachlorotoluene which has a CH_3 doublet with a separation of only 0.63 ± 2 Hz.[6] The Cl atoms are obviously not so effective in double-bond localization as the Br atoms.

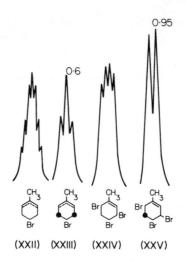

Figure 16. Comparison of the CH_3 signals in the n.m.r. spectra of
toluene and its bromine derivatives

The above results cannot be explained by the assumption of
Kekulé structures with three double bonds but can easily be
caused by the existence of only one true mobile double bond in any
benzenoid ring. This would be placed in the lowest orbital of the
three π orbitals and could be repelled by the influence of bromine
or chlorine or attracted by the CO groups in quinones.[4]

References

1. E. Clar and A. Mullen, *Tetrahedron*, **24**, 6719 (1968).
2. E. Clar and C. C. Mackay, *Tetrahedron Letters*, **11**, 871 (1970).
3. P. M. E. Lewis, *Tetrahedron Letters*, **21**, 1859 (1971).
4. E. Clar and C. C. Mackay, *Tetrahedron*, **27**, 5943 (1971).
5. M. P. Williamson, R. J. Kostelnik and S. M. Castellano. *J. Chem. Phys.*, **49**, 2218 (1968).
6. H. Rottendorf and S. Sternshell, *Australian J. Chem.*, **17**, 1315 (1964).

CHAPTER 11

The Aromatic Sextet in *peri*-Condensed Hydrocarbons

Biphenyl, Fluorene, Biphenylene and Perylene

Biphenyl is correctly presented with two sextets as in formula **I**. The length of the bond between the two rings is 1.5 Å which is almost the value for a single bond. The interaction between the two rings is mainly limited to the excited state. It is not an aromatic conjugation.

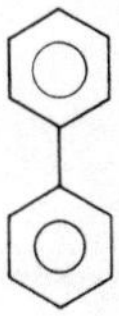

(I) 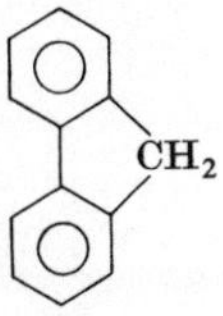   (II) 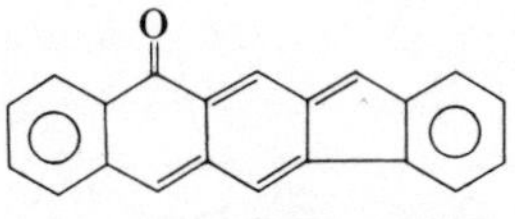   (III) 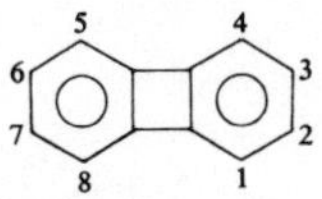   (IV)

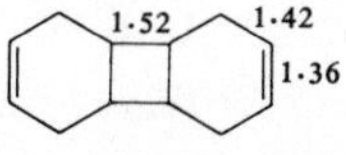

(V)

Fluorene (**II**) must be equally formulated with two sextets. There are no compounds known where the CH_2 is in one of the side rings, although this structure can be enforced in anthrone derivatives such as **III**.[1]

If biphenylene (**IV**) is written with two sextets then the two central bonds must be single bonds. Their length is indeed 1.52 Å. The double bond within the sextet appears to be mainly located in the 2,3- and 6,7-positions (**V**). These are the positions where chemical reactions take place, whilst in naphthalene the 1-position is favoured. Hydrogenation splits the molecule and gives biphenyl leaving the sextets untouched.[2]

The coupling constants derived from the n.m.r. spectrum are $J_{1.2} = 7.1$, $J_{2.3} = 8.1$ Hz confirming the double-bond location. However, the central four-membered ring is not completely void of π electrons. There are indications of a weak ring current in this ring.[3] There is also some degree of aromatic conjugation through this ring because an annellation effect can be recorded in the electronic spectra for the series **V**, **VI** and **VII** as shown by the shift of the β bands. This effect is bigger in the series **V**, **VIII** and **IX**. In accordance with this the latter hydrocarbons are more reactive than the linear ones.[4] This is in contrast to the acene series.

These shifts are small by comparison with the acene and phene series. There is a double-bond delocalization in the first series and the hydrocarbons become increasingly more stable. The members of the second series are more reactive and deeply coloured and there must be an increasing localization of double bond as shown in **VIII** and **IX**. In accordance with this benzobiphenylene (**VIII**) is oxidized to benzophenon-*o,o'*-dicarboxylic anhydride (**X**).[5]

The above hydrocarbons are not *peri*-condensed hydrocarbons, but biphenyl is the parent compound for these hydrocarbons from which they are derived by annellation, e.g. perylene.

Perylene can be considered to be built up from two naphthalene complexes connected by two single bonds (**XI**). This is a fairly correct description since the central bonds have a length of 1.50 Å.

(**XI**) (**XIa**) (**XII**) (**XIb**)

The other bond lengths are similar to the ones in naphthalene.[6] It is certainly not justifiable to write a sextet into the central ring. There are no indications for this from the measurement of the magnetic susceptibility[7] and the n.m.r. spectra do not support the assumption of a ring current in the central ring. The protons marked $H_{1,6,7,12}$ have a chemical shift of 8.05 p.p.m. which is equal to 805 Hz at 100 MHz. This is about 60 Hz at low field from the other protons $H_{3,4,9,10} = 757$ Hz and $H_{2,5,8,11} = 735$ Hz (**XIa, b**). This difference is only caused by the influence of the ring current in the other naphthalene complex, because the value 805 Hz for the "bay" protons is still considerably lower than for the bay protons in phenanthrene **XII** (marked H) which is 850 Hz.[8] This also proves the existence of an induced sextet in phenanthrene (**XII**) as marked by the arrows.

As in naphthalene there is no fixation of double bonds in perylene. They are partly located as shown by the dotted lines in the formulae (**XIa**). This can be demonstrated by the benzylic coupling in 3-methyl- and 3,9-dimethylperylene (**XIII**). Both hydrocarbons have CH_3 doublets with a separation of 0.6 Hz.[8]

The double-bond character in the bay positions can be shown by the diene synthesis with maleic anhydride. This yields 1,12-benzoperylenedicarboxylic anhydride (**XV**) via the intermediate product (**XIV**). **XIV** is so unstable and decomposes so readily into its constituents that a dehydrogenating agent like chloranil must be applied in the condensation. Thus the ring system is stabilized by the formation of a third sextet in the benzoperylene complex in **XV**.[9]

(XIII)   (XI)

(XIV)   (XV)

The perylene character is retained in the linear benzologues
XVI, **XVII** and **XVIII**. The frequencies of the bay protons are
given directly in the formula in Hz from TMS at 100 MHz and the
wavelength of the p and β bands underneath the formula.[10]

(XI)   (XVI)   (XVII)   (XVIII)

$\lambda_p = 4340 \xrightarrow{+690} 5030 \xrightarrow{+430} 5460 \xrightarrow{+610} 6070$ Å

$\lambda_\beta = 2510 \xrightarrow{+250} 2760 \xrightarrow{+140} 2900 \xrightarrow{+120} 3020$ Å

The electronic spectra are rather strongly influenced by steric
factors, i.e. overcrowding. Otherwise the annellation effect is
similar to that in the acene series. One would expect this since the
naphthalene complexes of perylene are replaced by anthracene
and tetracene complexes. This is only possible if the central ring
does not contain π electrons which could conjugate the two
moieties.

This independence is confirmed by the bay protons in the n.m.r.
spectra. They are at 805, 803, 810 and 816 Hz at 100 MHz. Their

position in perylene (**XI**) is not significantly altered in the benzologues in spite of the higher degree of overcrowding in **XVII** and
XVIII. It appears that the π electrons migrate only in the direction
given by the arrows. The perylene character is also retained with
respect to the coupling constants which remain about $J = 8$ for the
α,β position and $J = 7$ for the β,β position in **XVI** and **XVII**.[10]

Quite different annellation effects can be observed in the angular
series: **XI**, **XIX**, **XX** and **XXI**. Figures are given for the bay protons
of the n.m.r. spectra and the wavelengths of the p and β bands in
the electronic spectra:

(XI)		(XIX)		(XX)		(XXI)	
$\lambda_p = 4340$	$\xrightarrow{+60}$	4400	$\xrightarrow{-5}$	4335	$\xrightarrow{-85}$	4250	Å
$\lambda_\beta = 2510$	$\xrightarrow{+510}$	3020	$\xrightarrow{+15}$	3035	$\xrightarrow{+235}$	3270	Å

The p bands shift about in the same way as in passing from naphthalene to phenanthrene and benzophenanthrene. However, the
shifts are far too big (+510) for such a comparison, considering
XIX and **XX** as composed of two phenanthrene complexes connected by two single bonds. If perylene is built up from two
naphthalene complexes (β band at 2210 Å) there is a shift of 300 Å
for two single bonds or 150 Å for one single bond. This is a quite
reasonable procedure and has proved to give good results (see
p. 59). However, a shift of +510 Å for two single bonds rather
indicates that aromatic conjugation in the middle ring is involved
in **XIX**, **XX** and **XXI**. One could imagine that the arrows extend
into the middle ring. How this could be done involving a refinement
of the formulae will be shown in the last chapter.

For the time being aromatic conjugation in the middle ring is
also indicated by the strong shifts of the bay protons which are
now in the range of the bay protons of phenanthrene, i.e. 833, 854,
837, 840 and 872 Hz ex TMS at 100 MHz.[10]

(XIX)   (XXII)

The double bonds in perylene (**XI**) which are delocalized as indicated by the dotted line become fixed in dibenzoperylene (**XIX**). Thus a true *cis*-butadiene within an aromatic framework is formed. This reacts under the mildest conditions even in the absence of a dehydrogenating agent quantitatively with maleic anhydride to yield the adduct **XXII**.[9] The same condensation with tetrabenzoperylene leads to a double addition.[11] It should be remembered that perylene needs a dehydrogenating agent for this reaction because the double bonds are not fixed (see p. 60).

References

1. N. P. Buu-Hoi, C. M. and P. Jacquignon, *Bull. Soc. Chim.*, **1970**, 3958.
2. W. C. Lothrop, *J. Amer. Chem. Soc.*, **63**, 1187 (1941); W. Baker, M. P. V. Boarland and J. F. W. McOmie, *J. Chem. Soc.*, **1954**,1476.
3. A. R. Katritzky and R. E. Reavill, *Rec. Trav. Chim. Pays-Bas*, **83**, 1230 (1964); A. J. Jones and D. M. Grant, *Chem. Commun.*, **1968**, 1670.
4. M. P. Cava and D. R. Napier, *J. Amer. Chem. Soc.*, **79**, 1701 (1956); **80**, 2255 (1958); M. P. Cava and J. F. Stucker, *J. Amer. Chem. Soc.*, **77**, 6022 (1955).
5. M. P. Cava and J. F. Stucker, *J. Amer. Chem. Soc.*, **77**, 6022 (1955).
6. D. M. Donaldson, J. M. Robertson and J. G. White, *Proc. Roy. Soc. A*, **220**, 311 (1953).
7. H. Akamatu and Y. Matsunaga, *Bull. Chem. Soc. Japan*, **26**, 364 (1953); **29**, 800 (1956).
8. E. Clar, A. Mullen and Ü. Sanigök, *Tetrahedron*, **24**, 2817 (1968).
9. E. Clar and M. Zander, *J. Chem. Soc.* **1958**, 1861.
10. E. Clay and C. C. Mackay, Future communication.
11. E. Clar and B. A. McAndrew, *Tetrahedron*, **28**, 1137 (1972).

CHAPTER 12

Benzoperylene, Coronene and Superaromaticity

In benzoperylene (**I**) there are three sextets and two fixed double bonds. There will be aromaticity in the middle ring and induced sextets in the rings with the fixed double bonds. This multiplicity of sextet interaction is not symbolized by arrows in formula **I**. The n.m.r. spectrum of benzoperylene is clear and suitable for direct inspection (Figure 17).

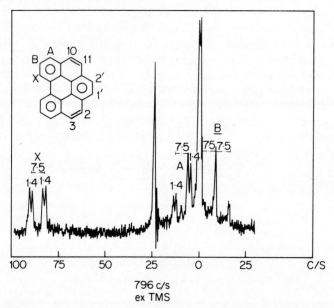

Figure 17. N.m.r. spectrum of 1,12-benzoperylene in CS_2 at 100 MHz. Protons centred at A, 803.0; B, 788.0; X, 881.6; $H_{3,10}$ 796.0; $H_{2,11}$, 795.0; $H_{1',2'}$ 819.0 Hz ex TMS

The position of the low-field protons X at 881.6 Hz shows clearly the influence of a ring current in the central ring. This is quite different from perylene where the corresponding value is at 805 Hz. The coupling constants $J_{AB} = J_{BX} = 7.5$ Hz are equal.

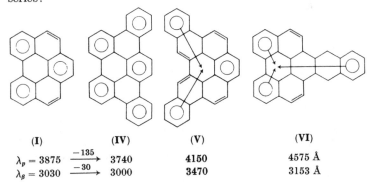

| (I) | (Ia) | (II) | (III) |

This shows that the double bond is equally distributed between two positions as marked by dotted lines in **Ia**. These are in fact two three-centred bonds. This causes the CH_3 doublet in the n.m.r. spectrum of **II** to have a separation of only 0.6 Hz.[1] The same applies to 4-methyl-1,12-benzoperylene. The two CH_3 groups in dimethylbenzoperylene (**III**) are attached to fixed double bonds. Therefore the doublet separation has the full value of 1.1 Hz.[2] These results are quite different from the ones obtained in perylene and can be explained only by the sextet formulae. Another difference from perylene is the strong red shift of the β band in the u.v. spectrum of benzoperylene which is 520 Å in passing from perylene to benzoperylene. This must be related to the intimate aromatic conjugation provided by the sextet connecting the two moieties in perylene.

There are the following annellation effects in the benzoperylene series:

| (I) | (IV) | (V) | (VI) |

$\lambda_p = 3875 \xrightarrow{-135} 3740 \qquad 4150 \qquad 4575$ Å

$\lambda_\beta = 3030 \xrightarrow{-30} 3000 \qquad 3470 \qquad 3153$ Å

This confirms the well-established rule that a ring fused to a fixed double bond causes either no shift or a weak one towards shorter wavelength. Considerable shifts to the red are observed if annellation takes place onto a sextet, i.e. in passing to **V** and **VI**.

The following series give interesting data about the n.m.r. spectra.[3]

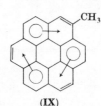

The coupling constants are recorded on the upper half and the chemical shifts on the lower half of the molecule. The most striking feature is the migration of the bay protons to higher fields with increasing annellation. This proves that the sextet in the middle is gradually diluted along the direction of the arrow and the ring current acting on the bay protons is reduced from 881.6 in **I** to 871 in **VII** and 867 Hz in **VI**.

If an ethylene bridge is fused to benzoperylene (**I**) to form coronene (**VIII**) a profound change takes place in the n.m.r. spectrum. All the protons shift to the position of the low-field protons in benzoperylene to form a single band at 872 Hz. If the shift of the latter is produced by a ring current in the centre then a similar ring current must extend over the whole coronene molecule. It is easy to see how this can happen. If the three sextets in coronene migrate into the neighbouring rings a further ring current, originating from the rotating sextets, must be formed as indicated by the arrows.[2]

Without this ring current, coronene (**VIII**) would have reactive double bonds like phenanthrene. This is not so. Coronene is a pale

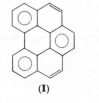

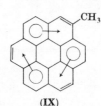

yellow, very stable hydrocarbon which does not dissolve in concentrated sulphuric acid and shows a yellow phosphorescence of very long life at low temperature in solid solution. These are properties which are reserved for fully benzenoid hydrocarbons (see Chapter 6). The position of the CH_3 group in methylcoronene (**IX**) is at unusually low field at 324 Hz at 100 MHz = τ 6.76. This signal is a doublet with a separation of 0.9 Hz which is a distinctly lower value than in phenanthrene.[4] The stability produced by the sextet ring current in coronene may be called "superaromaticity".

It makes such a strong contribution to the stability of the molecule that it is retained in benzocoronene (**X**) although there is one sextet less in one phase of the ring current **Xa** ⇌ **Xb**:[2]

(**Xa**)   (**Xb**)

(**XIa**)   (**XIb**)

The figures in **Xa** are chemical shifts in Hz from TMS at 100 MHz in CS_2. A coupling constant is given ($J = 9.0$) in **Xb**. There must also be superaromaticity in **XIa** ⇌ **XIb** with the same mechan-

(**XII**)   (**XIII**) ism as in benzocoronene. The same applies to naphthocoronene. But there can be no superaromaticity in 1.2,5.6-dibenzocoronene (**XII**). A ring current would enforce the participation of structure **XIII** which has two sextets less than **XII**. The measurements make it obvious that superaromaticity does not compensate for the loss of two sextets.

There is also no superaromaticity in hexabenzocoronene (**XIV**). If the sextets move between neighbouring rings (**XIVa⇌XIVb**) all rings achieve sextets although no circular movement is created:

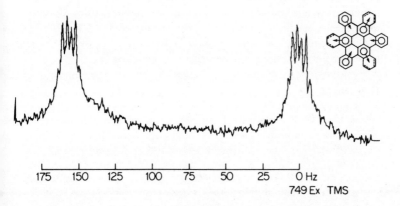

(**XIVa**)　　　　　(**XIVb**)

The n.m.r. spectrum of hexabenzocoronene (**XIV**) is presented in Figure 18. It is a remarkably clear spectrum which still belongs to the naphthalene-type spectrum in spite of the very wide separation of the α and β protons of 156 Hz.

175　150　125　100　75　50　25　0 Hz
749 Ex TMS

Figure 18.　N.m.r. spectrum of hexabenzocoronene in hexamethyl-benzene at 179° at 100 MHz

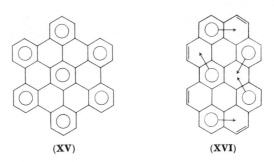

(XV)   (XVI)

The hexa-*peri*-benzocoronene (**XV**) does not require a super ring current. It is a yellow fully benzenoid hydrocarbon of extreme stability. In the mass spectroscope it forms a three-fold negative ion without fragmentation. It does not dissolve in concentrated sulphuric acid and shows an orange-red phosphorescence of very long life in solid solution at low temperature.[5]

The circobiphenyl **XVI** takes its name from the ten rings which surround the central biphenyl system, just as in coronene a benzene ring is surrounded by six benzene rings. This pale yellowish-green hydrocarbon has the properties of a fully benzenoid hydrocarbon. It is insoluble in concentrated sulphuric acid and has a red phosphorescence of very long life in solid solution at low temperature. It is for this reason that it must be formulated with two counter-running ring currents as indicated by the arrows. A high degree of superaromaticity must be derived from the n.m.r. spectrum in hexamethylbenzene at 200°. It contains a singlet at $1.35\,\tau$ Hz and a quartet originating from the bay proton centred at 0.08 and $0.14\,\tau$ Hz from the neighbouring protons. These are the lowest values recorded for an aromatic benzenoid hydrocarbon.[6]

References

1. Unpublished results.
2. E. Clar, Ü. Sanigök and M. Zander, *Tetrahedron*, **24**, 2817 (1968).
3. Future communication.
4. E. Clar, B. A. McAndrew and M. Zander, *Tetrahedron*, **23**, 985 (1967).
5. E. Clar and C. T. Ironside, *Proc. Chem. Soc.*, **1958**, 150; E. Clar, C. T. Ironside and M. Zander, *J. Chem. Soc.* **1959**, 142.
6. E. Clar and C. C. Mackay, Future communication.

CHAPTER 13

The Bisanthene Series

The fusion of two anthracene complexes gives the blue hydro-carbon bisanthene (**I**) (meso-naphthodianthrene). The β band in its electronic spectrum is at 3120 Å. This is at +605 Å higher than the β band of anthracene. There are three bonds connecting the two moieties, thus each bond accounts for a shift of +200 Å. These bonds must therefore be single bonds and there is no justi-fication to consider the two central rings as something other than empty rings. The two sextets move only in the direction of the arrows. Measurements of the magnetic susceptibility confirms the isolation of the moieties.[1]

If by benzeneogenic diene synthesis one ring is annellated in such a way that a new sextet is formed in benzobisanthene (**II**) an intimate conjugation between the two anthracene complexes

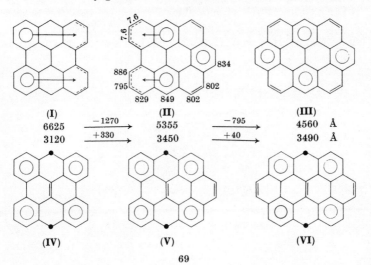

	(I)		**(II)**		**(III)**	
	6625	$\xrightarrow{-1270}$	5355	$\xrightarrow{-795}$	4560	Å
	3120	$\xrightarrow{+330}$	3450	$\xrightarrow{+40}$	3490	Å

(IV)   **(V)**   **(VI)**

is established. This results in a strong shift of +330 Å of the β band towards the red and an interesting even bigger shift of the p bands to shorter wavelength of −1270 Å.[2]

The n.m.r. values for benzobisanthene (**II**) are recorded in the lower half in Hz ex TMS at 100 MHz and the coupling constant in the upper half of the molecule. The remarkable feature amongst these figures is that the bay protons at 886 Hz have shifted very little to lower field by comparison with 1.12-benzoperylene (881.6 Hz) because no new sextet entered the central part of the molecule.

Once the aromatic conjugation has been established in benzo-bisanthene (**II**) a further sextet in passing to ovalene (**III**) produces only a small shift of +40 Å. The same relation exists between 1,12-benzoperylene and coronene. However, the p bands show another big shift of −795 Å to shorter wavelength.[2] In order to understand this one must consider the excited state of the three hydrocarbons in which two π electrons are localized as marked with points in **IV**, **V** and **VI**. The transition from **I** to **IV** increases the number of sextets from two to four, between **II** and **V** there is an increase from three to four sextets and between **III** and **VI** no change in the number of sextets takes place. It is obvious that an increase in the number of sextets facilitates the transition. Hence the strong shift of the p bands to the violet in going from **I** to **II** and **III**.

The number of the sextets plays a great role in the electronic spectra of the benzobisanthenes **VII** and **VIII**. There is less aromatic conjugation in **VIII** than in **VII**. Therefore the β bands

(**VII**)   (**VIII**)

$\lambda_p = 4170$ $\xrightarrow{+650}$ 4820 Å

$\lambda_\beta = 3580$ $\xrightarrow{-290}$ 3290 Å

shift towards shorter wavelength and the p bands to the red.[3]
These changes are quite similar to those in the preceding series.

In passing from the orange-yellow ovalene (**III**) to the dark-red
circumanthracene (**IX**) there is no change in the number of sextets
and all the absorption bands must shift to the red as in acene series.[4]

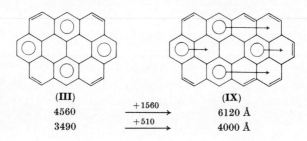

	(**III**)		(**IX**)
	4560	$\xrightarrow{+1560}$	6120 Å
	3490	$\xrightarrow{+510}$	4000 Å

References

1. H. Akamatu and Y. Matsunaga, *Bull. Chem. Soc. Japan*, **26**, 366 (1953).
2. E. Clar, *Polycyclic Hydrocarbons*, *I*, Academic Press, New York, 1964, *I*, p. 42; *II*, pp. 98, 100, 103.
3. E. Clar, G. S. Fell, C. T. Ironside and A. Balsillie, *Tetrahedron*, **10**, 26 (1960).
4. E. Clar, W. Kelly, J. M. Robertson and M. G. Rossmann, *J. Chem. Soc.*, **1956**, 3878.

CHAPTER 14

The Pyrene Series

In formula **I** pyrene has two sextets and two fixed double bonds. The latter must cause very small shifts in passing from **I** to **II** and **III** which is the case. Moreover, there is an element of electronic asymmetry as indicated by the arrows. This is not strikingly obvious in the series **I**, **II** and **III** because the rings are fused to fixed double bonds, but it becomes very obvious in going from biphenyl (**IV**) to phenanthrene (**V**) and pyrene (**I**).[1]

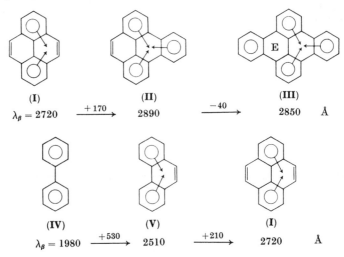

(**I**)   (**II**)   (**III**)

$\lambda_\beta = 2720$ $\xrightarrow{+170}$ 2890 $\xrightarrow{-40}$ 2850   Å

(**IV**)   (**V**)   (**I**)

$\lambda_\beta = 1980$ $\xrightarrow{+530}$ 2510 $\xrightarrow{+210}$ 2720   Å

The asymmetry in dibenzopyrene can be easily demonstrated by a comparison with dibenzophenalene in Figure 19. Both hydrocarbons have absorption bands at about the same position because in both the aromatic complex, as marked by shadow, is the same as in phenylphenanthrene, the third branch in dibenzopyrene

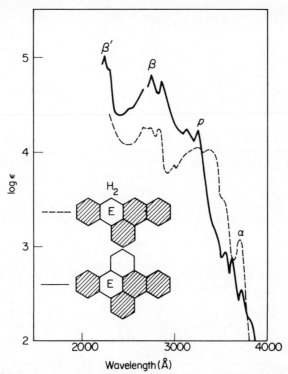

Figure 19. Absorption spectra. - - - -, Dibenzophenalene;
———, dibenzopyrene

being ineffective. There is no need to prove the aromatic conjugation in dibenzophenalene because the CH_2 is visible in the n.m.r. spectrum.[2]

The asymmetric annellation effect in the pyrene series can be readily demonstrated beginning with naphthopyrene. This has one fixed double bond. If first one and then a second ring are annellated the absorption spectra show no strong shift towards the red as would be required by aromatic conjugation but very small shifts towards shorter wavelength.

This can be even more drastically presented beginning with anthraceno–pyrene passing to benzoanthraceno–pyrene (Figure 21) and then to naphthoanthracenopyrene and dianthracenopyrene (Figure 22). Only minor changes can be observed. If

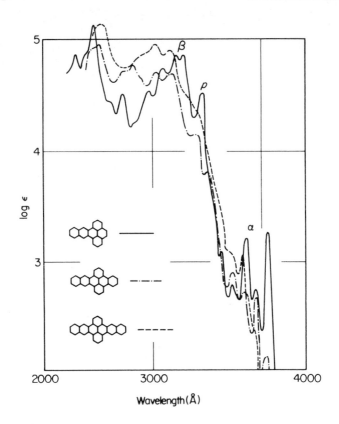

Figure 20. Absorption spectra of naphthopyrene, benzonaphtho-
pyrene and dinaphthopyrene

in dianthracenopyrene the eight linear rings were in aromatic
conjugation forming an octacene system there should be a red
shift of several thousands of Å and the hydrocarbons would be
very unstable. In fact they are orange and very stable, because
aromatic conjugation with the central biphenyl system is restricted
to only one moiety of the molecule.[1,2]

Annellation to the two sextets in pyrene produces an annellation
series which is similar to the acene series with the difference that
two rings bring about the same effect as one ring in the acene
series. The number of sextets remains two. The annellation effect

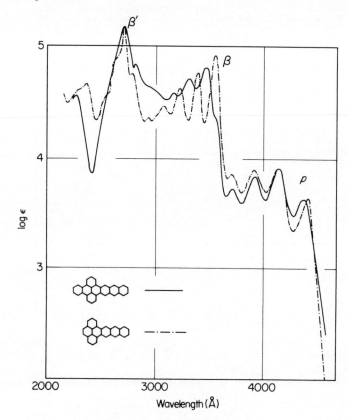

Figure 21. Absorption spectra of anthracenopyrene and benzo-
anthracenopyrene

of the β bands is somewhat irregular. This results from the asymmetric members of the series. If only symmetric and asymmetric members are compared the differences become more constant.

In contrast to the above series there are two different kinds of aromatic conjugation possible in the following hydrocarbons. The dibenzopyrene can be considered a pentaphene derivative (**VIb**) or a picene derivative (**VIa**). In both cases the aromatic conjugation is marked by arrows and empty rings "E". Each kind of conjugation produces its own β band. One must assume that during the time of light absorption they are different species. If

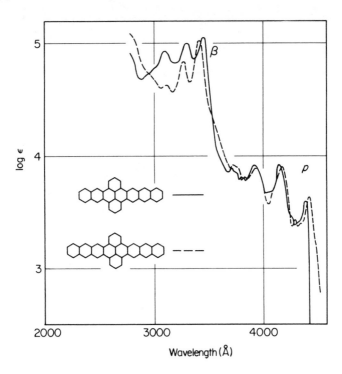

Figure 22. Absorption spectra of naphthoanthracenopyrene and dianthracenopyrene

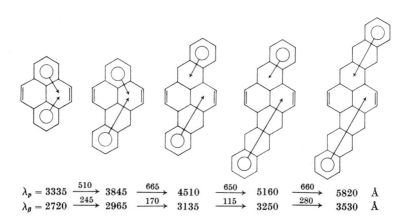

$$\lambda_p = 3335 \xrightarrow{510} 3845 \xrightarrow{665} 4510 \xrightarrow{650} 5160 \xrightarrow{660} 5820 \ \text{Å}$$
$$\lambda_\beta = 2720 \xrightarrow{245} 2965 \xrightarrow{170} 3135 \xrightarrow{115} 3250 \xrightarrow{280} 3530 \ \text{Å}$$

rings are fused to the fixed double bonds in dibenzopyrene small shifts are observed as would be expected. In the tribenzopyrenes the picene structure is stabilized in **VII** and the pentaphene structure in **VIII**. The fixed double bond in **VIII** was proved by the benzylic coupling in the n.m.r. spectrum of the CH_3 derivative (**VIII**, R = CH_3) which amounts to 1.1 Hz. The annellation of two

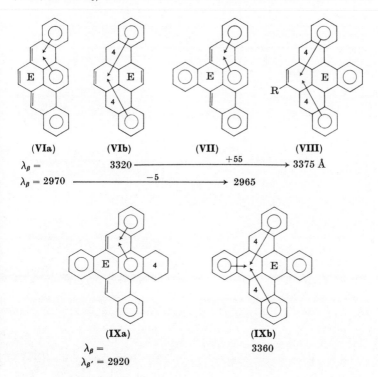

| (VIa) | (VIb) | (VII) | (VIII) |

$\lambda_\beta =$ 3320 $\xrightarrow{\;\;+55\;\;}$ 3375 Å

$\lambda_\beta = 2970 \xrightarrow{\;\;-5\;\;} 2965$

| (IXa) | (IXb) |

$\lambda_\beta =$ 3360

$\lambda_{\beta'} = 2920$

rings to the fixed double bonds in dibenzopyrene results in the formation of tetrabenzopyrene. Two β bands originating from the two structures **IXa** and **IXb** can be observed; the band at 2920 Å belonging to the picene type (**IXa**) and the band at 3360 Å belonging to the pentaphene type (**IXb**).[2]

The n.m.r. spectra of methylpyrenes give full support to the assumption that pyrene consists of two sextets and two fixed double bonds.

The methylpyrene **X** has a CH_3 doublet with a separation of 1.1

Hz and the dimethylpyrene **XI** a doublet with a separation of
1.1 Hz, both values indicating double-bond fixation.

Hydrogenation of **XI** to the naphthalene derivative **XII** transforms the doublet into a sharp singlet. This does also prove that
there is no true double-bond character between β,β positions in the

(**X**)   (**XI**)   (**XII**)   (**XIII**)

(**XIV**)   (**XV**)   (**XVI**)   (**XVII**)

central naphthalene complex in **XII**. The dimethylpyrene **XIII**
has a triplet with a separation of 0.55 Hz. This can only originate
from three-centred bonds as marked by dashed lines. (The sextets
are not marked in this and the following formulae.) The monomethyl derivative also has a CH_3 triplet.

There are two three-centred bonds in tetramethylpyrene (**XIV**)
forming doublets with a separation of 0.5 Hz. However, the
dimethylpyrene **XV** has a CH_3 singlet as has the corresponding
monomethylpyrene. This is somewhat surprising and one must
assume that the CH_3 groups repel the double bond.[3] However this
is supported by the result of the Vilsmeyer synthesis applied to
monomethylpyrene (**XVI**) which forms **XVII**.[4] This is the only
reaction in which the second substituent enters the same ring in a
pyrene complex. The distribution of double bonds and three-
centred bonds in **XVI** explains the formation of **XVII**.

There are some very interesting coupling reactions of hydroxypyrenes with diazo compounds which go exactly parallel with the benzylic coupling in methylpyrenes.

(XVIII)   (XIX)   (XX)

(XXI)

Hydroxypyrene (**XVIII**) couples readily with diazo compounds as indicated by the arrow. This is in line with the appearance of CH_3 triplets in **XIII** which have been correlated to the three-centred bonds. The hydroxypyrene **XIX** also couples readily with diazo compounds. In agreement with this, fixed double bonds were found adjacent to the dimethylpyrene **XI** and the corresponding monomethyl derivative. No diazo coupling can be observed with the hydroxypyrene **XX**.[5] In accordance with this the CH_3 groups in the dimethylpyrene **XV** and in the monomethyl derivative do not form doublets in their n.m.r. spectra because there appears to be no true double bond between CH_3 and the proton in the *ortho* position. There is thus a complete analogy between doublet formation in methylpyrenes and diazo coupling with hydroxypyrenes.

A note of warning should be given. As can be seen from the shifted double bond on dimethylpyrene (**XV**) it may be misleading to draw conclusions for the parent hydrocarbon from measurements of a methyl derivative. The CH_3 group can cause changes in the electronic fine structure of an aromatic system and several investigations are needed to draw firm conclusions. As a result of the above observations a distribution of double bonds and three-centred bonds as presented in formula **XXI** is the most likely in pyrene. It shows a mobile (three-centred) bond within the inherent sextets and fixed double bonds outside these sextets.

References

1. E. Clar, J. F. Guye-Vuillème, A. McCallum and I. A. Macpherson, *Tetrahedron*, **19**, 2185 (1963).
2. E. Clar, B. A. McAndrew and J. F. Stephen, *Tetrahedron*, **26**, 5465 (1970).
3. E. Clar, B. A. McAndrew and Ü. Sanigök, *Tetrahedron*, **26**, 2099 (1970).
4. M. de Clerq and R. Martin, *Bull. Soc. Chim. Belg.*, **64**, 367 (1955).
5. H. Vollmann, H. Becker, M. Corell and H. Streeck, *Liebigs Ann.*, **531**, 33, 59, 62, 72 (1937).

CHAPTER 15

The Anthanthrene Series

Anthanthrene is related to pyrene and could be considered a benzologue of pyrene. According to formula **I** it has two sextets which can move within the molecule as indicated by the arrows. The annellation effect is shown in the series: **I**, **II** and **III**. The number of sextets is increased in passing from **I** to dibenzanthanthrene **II**. There is no change in going from **I** to dibenzanthanthrene **III** and bigger shifts toward the red are observed.

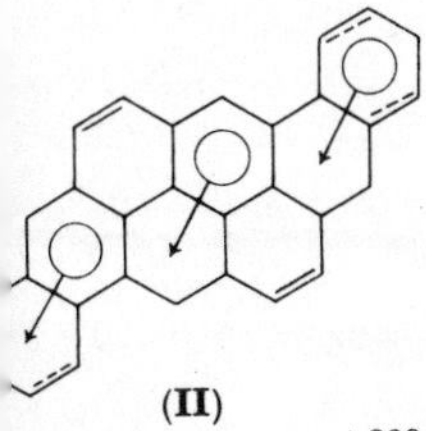

(**II**)

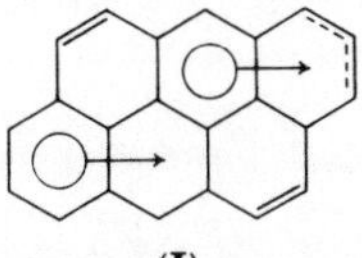

p = 4620	⟵ +290	4330	+560 ⟶	5890 Å	
β = 3540	⟵ +440	3100	+350 ⟶	3450 Å	

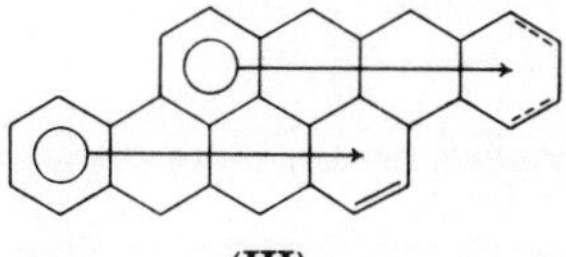

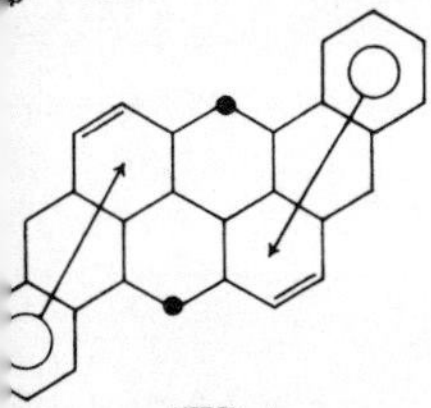

(**I**)   (**III**)

(**IV**) 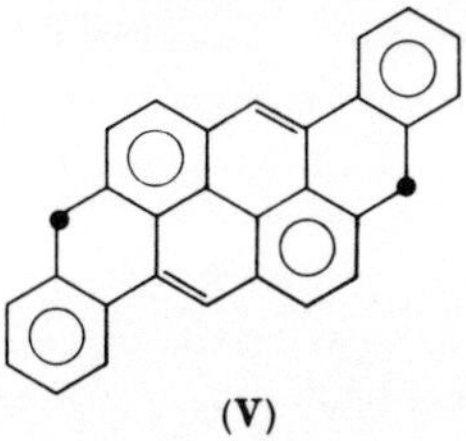   (**V**) 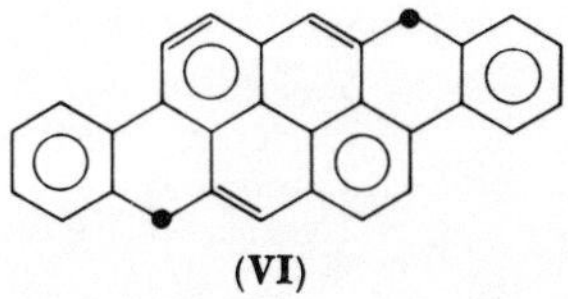   (**VI**)

There are two excited states for pyranthrene (**II**), i.e. **IV** and **V**. Pyranthrene is oxidized to pyranthrone if derived from the structure **V** with four sextets and not from structure **IV** with two sextets. This shows the importance of sextets also in the excited state. An analogous isomeric excited state with four sextets is shown in formula **VI**.[1]

Reference

1. E. Clar, *Polycyclic Hydrocarbons*, *I*, Academic Press, New York, 1964, *I*, p. 46.

CHAPTER 16

Hydrocarbons Derived from *p*-Terphenyl

The Terrylene Series

Terrylene (**I**) is derived from *p*-terphenyl and therefore has three sextets which move in such a way as to form three naphthalene complexes. This makes the red-violet terrylene a naphthologue of perylene. In fact it shows the annellation effects of perylene. There are strong shifts towards the red with linear annellation in

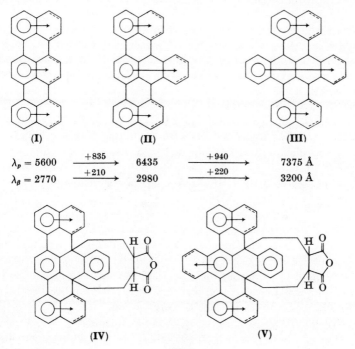

	(**I**)		(**II**)		(**III**)
$\lambda_p = 5600$	$\xrightarrow{+835}$	6435	$\xrightarrow{+940}$	7375 Å
$\lambda_\beta = 2770$	$\xrightarrow{+210}$	2980	$\xrightarrow{+220}$	3200 Å

(**IV**)

(**V**)

passing to the green-blue benzoterrylene (**II**) and the green dibenzoterrylene (**III**).[1]

The benzologues **II** and **III** react with maleic anhydride to form the endocyclic systems **IV** and **V**. As in the perylene series this reaction is restricted to hydrocarbons which have at least three linearly annellated rings, because only in these cases do the adducts have one sextet more than the corresponding hydrocarbons.

The adduct **IV** is colourless and **V** is yellow because the aromatic conjugation of the terrylene complex is interrupted and only the terphenyl and benzoterphenyl complex, respectively, are left intact.

As in the perylene series the angular benzologues in the terrylene series show little change by comparison with the parent hydrocarbons.

Thus the violet-red dibenzoterrylene (**VI**) has a p band which is only shifted slightly towards the violet whilst the β band shifts $+680$ Å towards the red. Owing to the two fixed double bonds it reacts readily with maleic anhydride and a trace of iodine to form the tetra adduct **VII**. This loses on sublimation two maleic anhydride complexes and goes over into the dianhydride **VIII** which has two sextets more than dibenzoterrylene **VI**. Decarboxylation of the dianhydride **VIII** yields the yellow fully benzenoid tetrabenzoterrylene **IX**. As expected it is a hydrocarbon of extreme stability.

The blue tribenzoterrylene **X** is very similar to benzoterrylene **II**, the p bands shifting by only -14 Å towards the violet. The relation between the two hydrocarbons is similar to that between naphthalene and phenanthrene. Hydrocarbon **X** also reacts readily with maleic anhydride to form a colourless endocyclic adduct **XI** which has one sextet more than **X**. A more powerful diene synthesis gives the dianhydride **XII** from which pentabenzoterrylene (**XIII**) can be obtained by decarboxylation. It is a red hydrocarbon quite different from the fully benzenoid hydrocarbon **IX**.

If two rings are cut from the fully benzenoid tetrabenzoterrylene (**IX**), dibenzoterrylene (**XIV**)[2] is obtained which is very little different. This confirms the rule that the third branch in any three-branched conjugation has very little influence because only two branches can be in aromatic conjugation.

The blue 1.9,5.10-di(*peri*-naphthylene) anthracene (**XV**) is very similar to the isomeric hydrocarbon **II**. It also adds maleic an-

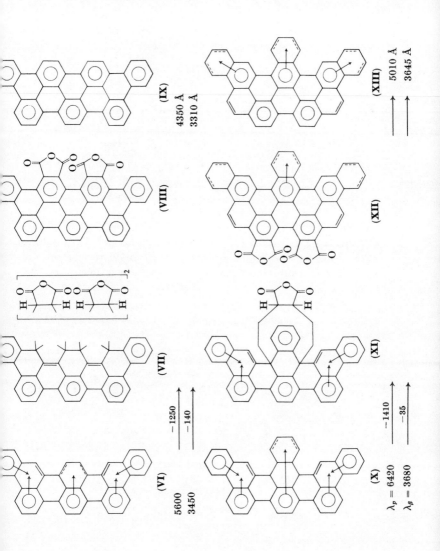

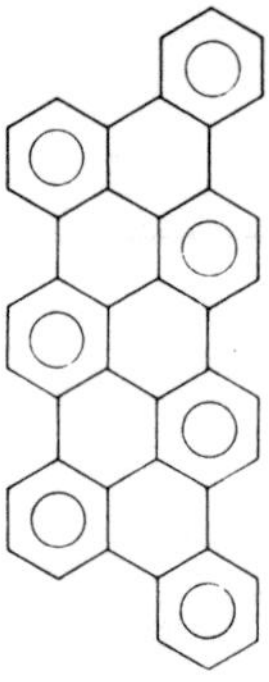

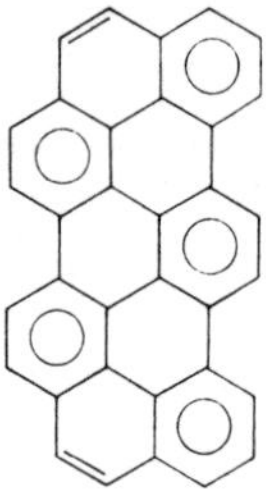

(IX)

$\lambda_p = 4350$ $\xrightarrow{\ -110\ }$ 4460 Å

$\lambda_\beta = 3310$ $\xrightarrow{\ 0\ }$ 3310 Å

(XIV)

hydride with the formation of the colourless endocyclic adduct **XVI**. A more powerful diene synthesis yields the dianhydride **XVII** which has two more sextets. The parent hydrocarbon will be considered in the next chapter. The above reactions prove the

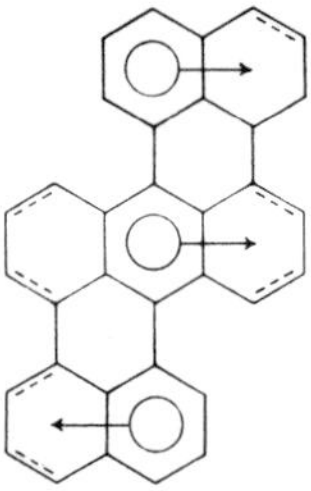

(XV)

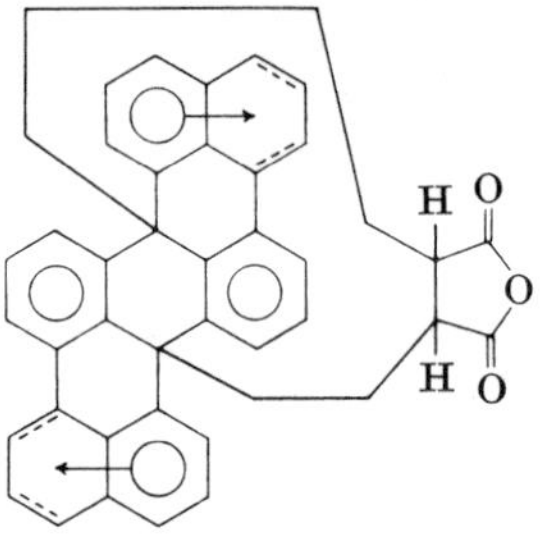

(XVI)

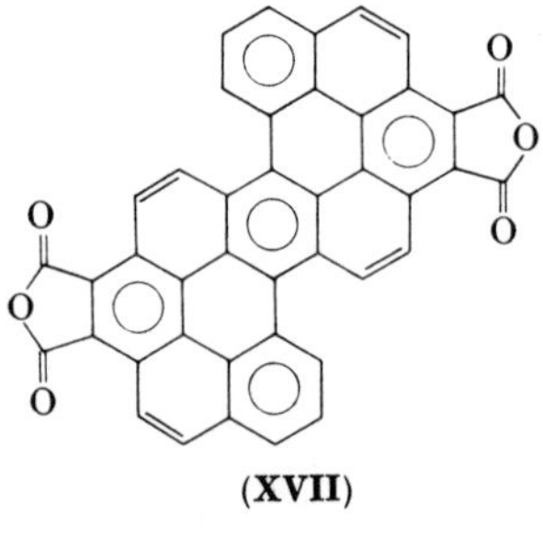

(XVII)

perylene and terrylene character of the hydrocarbons and by implication the correctness of the formulae written with aromatic sextets.[3]

The n.m.r. spectrum of benzoterrylene (**II**) confirms the similarity between the perylene and terrylene series. The bay protons are at 818 and 869 Hz ex TMS (in CS_2) at 100 MHz. This is very little different from the corresponding values in the perylene series. It does not indicate an appreciable ring current in the central ring between the naphthalene and anthracene complexes.

(II)

References
1. E. Clar and A. Guzzi, *Ber. Deut. Chem. Ges.*, **65**, 1521 (1932); E. Clar, W. Kelly and J. W. Wright, *J. Chem. Soc.*, **1954**, 1108; E. Clar and W. Willicks, *Ber. Deut. Chem. Ges.*, **88**, 1205 (1955).
2. E. Clar and O. Kühn, *Liebigs Ann.*, **601**, 181 (1956).
3. E. Clar and A. Mullen, *Tetrahedron*, **27**, 5239 (1971).

CHAPTER 17

Hydrocarbons Derived from *m*-Terphenyl

2.3-*peri*-Naphthylenepyrene (**I**) has three sextets which indicate the relation of this hydrocarbon to *m*-terphenyl. The orange-yellow hydrocarbon has moreover one fixed double bond in the bay position which makes it very suitable for a benzenogenic diene synthesis. It does in fact react very readily with maleic anhydride and chloranil to form the anhydride **II**. Decarboxylation yields pyreno[1,3 : 10′,2′]pyrene (**III**). The fully benzenoid tetrabenzoanthanthrene (**IV**) is related to it by annellation of two rings to fixed double bonds. As a result of this the spectral shifts are very small but they are considerable in passing from **I** to **III**.

(I)	(II)	(III)	
			(IV)

$\lambda_p = 4700$ $\xrightarrow{-640}$ 4060 $\xrightarrow{-20}$ 4040 Å

$\lambda_\beta = 2940$ $\xrightarrow{+280}$ 3220 $\xrightarrow{+20}$ 3240 Å

The annellation effects are quite similar to those in the perylene and pyrene series. It is therefore justifiable to consider the hydrocarbon **I** as a combination of perylene and pyrene.[1]

Reference

1. E. Clar and O. Kühn, *Liebigs Ann.* **601**, 181 (1956); E. Clar and A. McCallum, *Tetrahedron*, **20**, 507 (1964).

CHAPTER 18

Hydrocarbons Derived from Peropyrene

Peropyrene (**I**) consists of three sextets and four fixed double bonds and is derived from terphenyl in the same way as pyrene is from biphenyl. The fixed double bonds become obvious by annellation. The shifts are small but slightly bigger than in the pyrene series. This is due to overcrowding in the dibenzo- and tetrabenzoperopyrene **II** and **III**, respectively.

(**I**)	(**II**)		(**III**)
$\lambda_p = 4435$	$\xrightarrow{+15}$ 4450	$\xrightarrow{+220}$	4670 Å
$\lambda_\beta = 3260$	$\xrightarrow{+220}$ 3480	$\xrightarrow{+50}$	3530 Å

There are some interesting n.m.r. data available for peropyrene (**I**) and dibenzoperopyrene (**II**) (in Hz from TMS at 100 MHz in CS_2). The most striking value has the bay proton at 915 Hz in peropyrene (**I**). This is at very low field, much lower than in any perylene or phene-type hydrocarbon. This can only be explained by the inherent sextet in the centre which the preceding hydrocarbons have not. The large ring current is a direct proof for the inherent sextet in the centre. The double bonds reveal their nature by the coupling constant of 9 Hz.

Similar values are observed in dibenzoperopyrene (**II**). The bay protons are at 905 Hz. This shows that a higher degree of over-

crowding is less influential than the central sextet. The coupling
constant for the double bond remains at 9 Hz and the coupling
constants for the upper and lower sextet ring protons are 7.5 Hz
as in pyrene.

Annellations to the sextet rings have bigger effects, as in the
pyrene series. Whilst peropyrene (**I**) is a yellow hydrocarbon,
dibenzoperopyrene (isoviolanthrene) (**IV**) is red. It is a strange and
probably significant coincidence that dibenzoperopyrene (di-
benzocoronene) **V** has almost the same absorption spectrum as
violanthrene (**IV**).

$$\lambda_p = 4435 \xrightarrow{\ +795\ } 5230 \xrightarrow{\ +10\ } 5240 \text{ Å}$$
$$\lambda_\beta = 3260 \xrightarrow{\ +380\ } 3640 \xrightarrow{\ -20\ } 3620 \text{ Å}$$

Whilst isoviolanthrene has a single sharp β band the bright red
violanthrene (**VI**) has two β bands (at 3820 and 3275 Å). The
analogy with the pyrene series is complete. In both cases the
centrosymmetric hydrocarbons have a single and the plansym-
metric hydrocarbons, such as **IV** and **VI** respectively, have two
β bands. The latter can be referred to two types of aromatic
conjugation. More n.m.r. data were obtained from orange-red
dinaphthoperopyrene (**VII**). The bay protons migrate to very low
field at 942 Hz (from TMS at 100 MHz in CS_2). The bay protons
attached to the sextets are at slightly higher field at 923 Hz. The
upper part of the formula **VII** contains the coupling constants.
These are 9 Hz for the double bonds and 7.5 Hz for the sextets
in bay positions. The other protons form singlets.

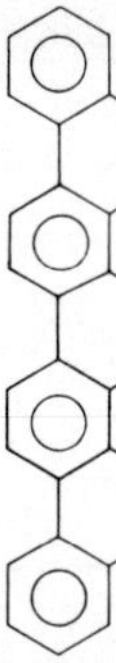

(VI) 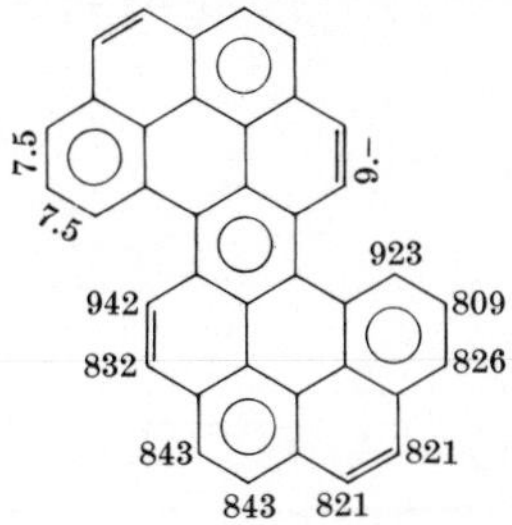   (VII) 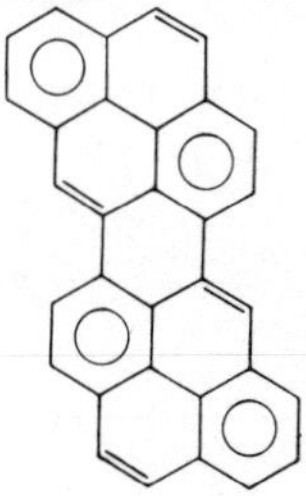   (VIII)

The hydrocarbon **VII** consists of two condensed 1,12-benzo-perylene systems and there is a remarkable similarity of the n.m.r. data between these two hydrocarbons.[1]

The orange dibenzoperopyrene **VIII** can be considered as *peri*-condensed from two pyrene molecules. It does not react very readily with maleic anhydride. However, this is the case with the isomeric hydrocarbon **IX**. This adds maleic anhydride without a dehydrogenating agent and forms the colourless adduct **X**. The reason for this reaction is that two double bonds are fixed in the *cis*-butadiene position. Dehydrogenation of **X** yields the anhydride **XI** and decarboxylation the golden-yellow tribenzoperopyrene (**XII**). This process is accompanied by big shifts to shorter wavelength as expected.[2] The very small effect of fusing rings to the fixed double bonds in **XII** has already been reported on p. 86.

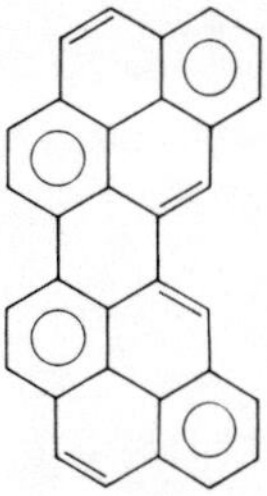

| (IX) | (X) | (XI) | (XII) |

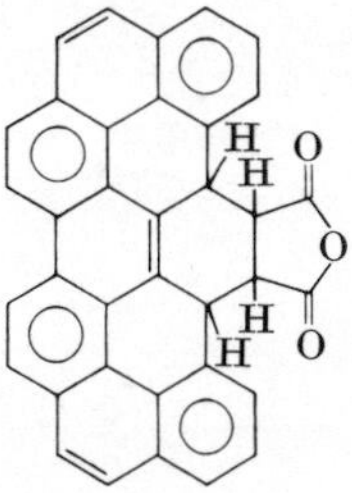

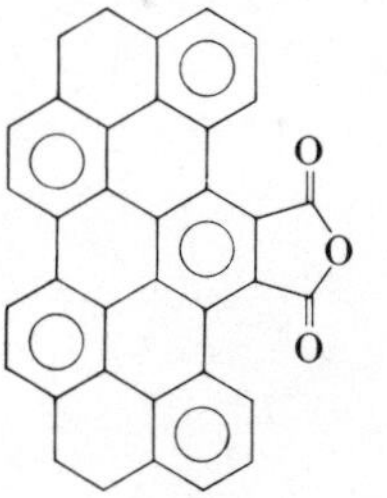

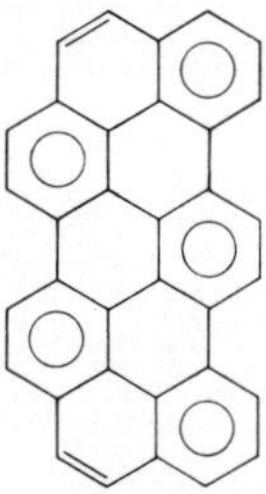

$\lambda_p = 5200 \xrightarrow{-740}$ $\longrightarrow$ 4460 Å

$\lambda_\beta = 3425 \xrightarrow{-115}$ $\longrightarrow$ 3310 Å

It is interesting to compare the two isomeric tetrabenzopero-pyrenes **XIII** and **XIV**. As usual the hydrocarbon with the smaller number of sextets absorbs at longer wavelength of the p band.[3]

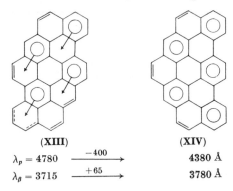

(**XIII**)   (**XIV**)

$\lambda_p = 4780$ $\xrightarrow{-400}$ 4380 Å

$\lambda_\beta = 3715$ $\xrightarrow{+65}$ 3780 Å

References

1. E. Clar and C. C. Mackay, Future communication.
2. E. Clar and O. Kühn, *Liebigs Ann.*, **601**, 181 (1956).
3. E. Clar and W. Kelly, *J. Chem. Soc.*, **1956**, 3875.

CHAPTER 19

The Quaterrylene Series

Quaterrylene is a green hydrocarbon of considerable stability. This stability is the result of the four sextets in formula **I**. A similar green hydrocarbon in the acene series such as hexacene is most reactive and unstable. Derived from quaterphenyl, quaterrylene (**I**) must be considered to consist of four naphthalene complexes connected by single bonds. This is confirmed by X-ray analysis. The bonds connecting the naphthalene complexes have a length of 1.529 and 1.527 Å. The other bonds have similar lengths to those in naphthalene.[1]

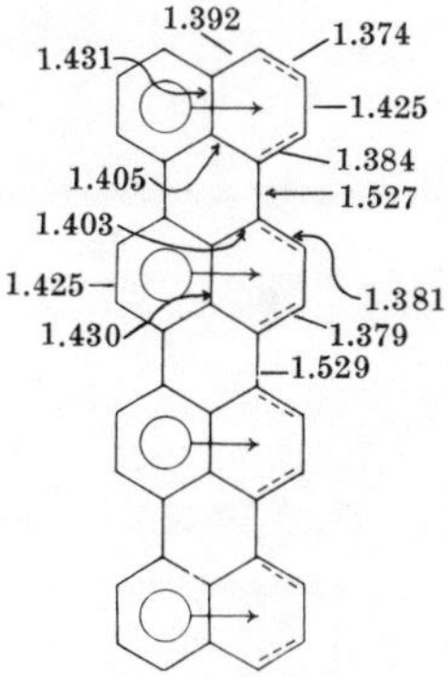

(**I**)

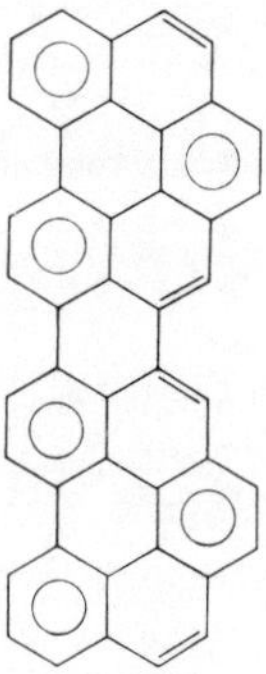

$$\lambda_p = 6700 \xrightarrow{-930} 5770 \text{ Å}$$

$$\lambda_\beta = \qquad\qquad\qquad 3440 \text{ Å}$$

(**II**)

Fusing two rings to quaterrylene (**I**) in going to dibenzoquaterrylene (**II**) brings an increase of two sextets. Therefore a big shift of 930 Å to shorter wavelength is recorded. Dibenzoquaterrylene (**II**)

93

is violet, can be sublimed at 500° and enters a diene synthesis with maleic anhydride and chloranil to form a brown anhydride.[2]

References

1. E. Clar and J. C. Speakman, *J. Chem. Soc.*, **1958**, 2492; H. N. Srivastava and J. C. Speakman, *Proc. Roy. Soc. A*, **257**, 477 (1960).
2. A. McCallum, Thesis, Glasgow (1963); E. Clar, *Polycyclic Hydrocarbons*, *II*, Academic Press, New York, 1964, p. 284.

CHAPTER 20

2.3,7.8-Di(*peri*-naphthylene)pyrene

In the same way as quaterrylene is derived from quaterphenyl the title hydrocarbon (**I**) is related to 3.3′-dibiphenyl. This system is shown by the sextets in formula **I**. Hydrocarbon **I** is dark violet and reacts readily with maleic anhydride and chloranil to form the dianhydride **II**. The latter gives the hydrocarbon **III** on dicarboxylation. It is yellow and has two more sextets than **I**. The shifts to shorter wavelength are very big in accordance with expectation.

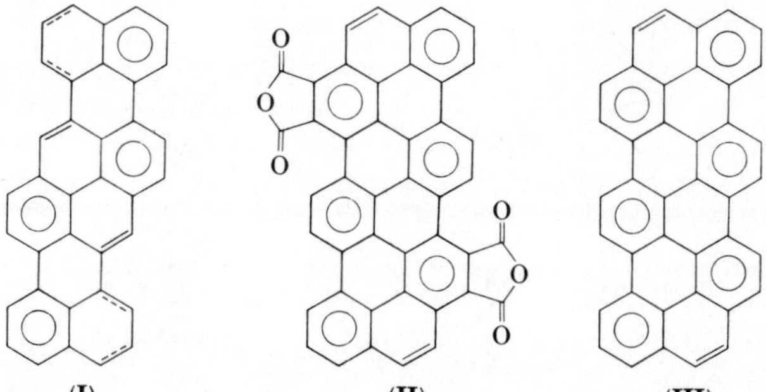

| (I) | (II) | (III) |

$$\lambda_p = 5750 \xrightarrow{-1050} 4700 \text{ Å}$$

$$\lambda_\beta = 3550 \xrightarrow{-150} 3400 \text{ Å}$$

Hydrocarbon **III** is yellow, very sparingly soluble, does not melt below 510° and is insoluble in concentrated sulphuric acid. These properties are in line with its highly benzenoid character.[1]

Reference

1. E. Clar and O. Kühn, *Liebigs Ann.*, **601**, 181 (1956).

CHAPTER 21

The Fluoranthene Series

According to formula **I** fluoranthene consists of a naphthalene and phenylene complex. There are α and β bands in the electronic absorption spectrum of fluoranthene. The β' bands at 2130 Å originate from the naphthanthalene complex and are caused by the absorption of light polarized in the molecular plane which enters in the direction of the short axis of the naphthalene complex or long axis of fluoranthene. The phenylene complex has little influence since only a shift of −80 Å can be recorded by comparison with the β band of naphthalene at 2210 Å.

	(I)		(II)		(III)
$\lambda_p = 3585$	$\xrightarrow{+695}$	4280	$\xrightarrow{+1430}$	5710 Å
$\lambda_\beta = 2130$	$\xrightarrow{+430}$	2560	$\xrightarrow{+365}$	2925 Å

	(IV)		(V)		(VI)
$\lambda_p = 4100$	$\xrightarrow{+750}$	4850	$\xrightarrow{+1240}$	6090 Å
$\lambda_\beta = 2175$	$\xrightarrow{+505}$	2680	$\xrightarrow{+370}$	3050 Å

+45   +120   +125

This remains so in passing to the annellated fluoranthenes **II** and **III** which are derived from anthracene and tetracene, respectively. Adding phenylene complexes to these hydrocarbons the shifts are 45 and 185 Å, respectively, for the β bands.

The annellation shifts are big but less regular than in the acene series. The shifts for the fusion of the second phenylene group are: 45, 120 and 125 Å. These are too small to account for anything other than the formation of single bonds.

(VII)   **(VIII)**   **(IX)**

$\lambda_\beta = 2210 \xrightarrow{+870} 3080 \xrightarrow{+210} 3290$ Å

(X)

     (XI)   **(XII)**

$\lambda_\beta = 2515 \xrightarrow{+825} 3340 \xrightarrow{+200} 3540$ Å

A different series starting from naphthalene (**VII**), passing first to benzofluoranthene (**VIII**) (shift +870 Å) and then to dinaphthylenenaphthalene (**IX**) (shift +210 Å) produces very asymmetric annellation effects.

The third series begins with anthracene (**X**). The first *peri*-naphthylene complex leads to naphthofluoranthene (**XI**) (shift +825 Å) and the second to dinaphthyleneanthracene (**XII**) (shift +200 Å). These asymmetries have been symbolized in the formula by arrows leading into the five-membered rings. This is not a complete description of the annellation effect in which the naphthalene complex also participates in order to fill the five-membered ring with electrons. The second annellation produces empty rings "E" which are formed by two single bonds.[1] The naphthalene complexes have been given fixed double bonds. This will be justified now.

Fluoranthene (**I**) like phenanthrene is oxidized to an *o*-quinone (**XIII**). In accordance with this 3-methylfluoranthene (**XIV**) has an n.m.r. spectrum with a CH_3 doublet with a separation of 1 Hz.

This is the same value as in 9-methylphenanthrene. It is therefore justifiable to write fixed double bonds adjacent to the CH_3 in both hydrocarbons.

(I)   (XIII)   (XIV)   (XV)

1-Methylfluoranthene (**XV**) has a CH_3 singlet which is little affected by decoupling the neighbouring aromatic proton. Therefore it cannot have a fixed double bond adjacent to the CH_3 group. This structure is confirmed by the n.m.r. spectrum of 1,3-dimethylfluoranthene (**XVI**) which has a CH_3 doublet (separation 1 Hz) and a singlet. 1,3,6-Trimethylfluoranthene (**XVII**) has two singlets and one doublet.

It has already been pointed out (p. 54) that neither 7,10-dimethylfluoranthene (**XVIII**) nor 8,9-dimethylfluoranthene (**XIX**) have CH_3 doublets, but sharp singlets. The rings with the CH_3 groups must have the fine structure of *p*- and *o*-xylene, respectively.[2]

(XVI)   (XVII)   (XVIII)   (XIX)

The most likely location of the double bond of the upper ring is adjacent to the five-membered ring as in **XX**. If rings are fused to fluoranthene (**XX**) leading to benzofluoranthene (**XXI**) and naphthofluoranthene (**XXII**) big shifts towards the red of the β bands are observed whilst there are only minor shifts of the *p* bands because the annellation is angular with regard to the fluoranthene complex. As far as extending an acene complex is concerned, the double bond in the upper ring remains in the same position. This annellation is therefore linear, as in the acene series. In addition a new p band (at 4820 Å) appears in naphthofluoranthene (**XII**) which is due to the anthracene complex. This appearance of new types of bands is bound to come if the annellation produces a new type (acene-type) which is not inherent in the parent type (fluoranthene-type). Quite different observations can be made in the series: **XX**, **XXIII** and **XXIV**. The p bands change

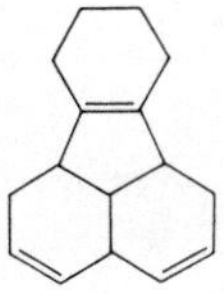

(**XX**) 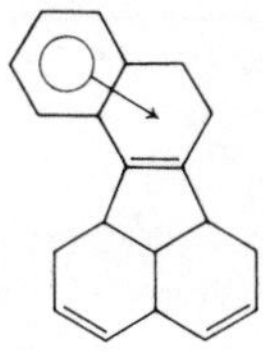   (**XXI**) 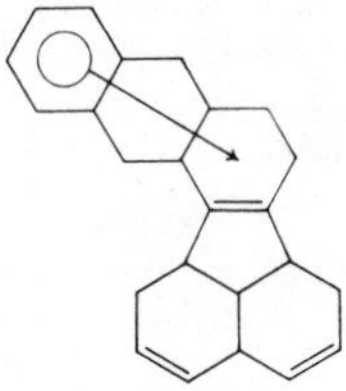   (**XXII**)

4820 Å (anthracene type)

$$\lambda_p = 3585 \xrightarrow{+245} 3830 \xrightarrow{+245} 4075 \text{ Å}$$
$$\lambda_\beta = 2370 \xrightarrow{+810} 3180 \xrightarrow{+410} 3590 \text{ Å}$$

between **XX** and **XXIII** from the fluorene-type p band to the acene-type producing very big sifts. Since there cannot be a location of a double bond in the naphthalene and anthracene complexes in the β,β position a change of the double bond in the upper fluoranthene ring must take place to the positions indicated by the dashed lines.

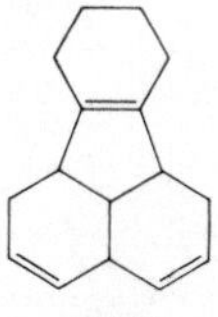

(**XX**) 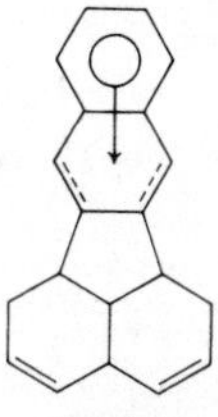   (**XXIII**) 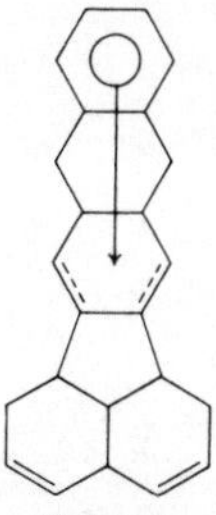   (**XXIV**)

$$\lambda_p = 3585 \xrightarrow{+415} 4000 \xrightarrow{+380} 4380 \text{ Å}$$
$$\lambda_\beta = 2370 \xrightarrow{+710} 3080 \xrightarrow{+260} 3340 \text{ Å}$$

In fusing a ring to the double bond in the 2,3-position in fluoranthene the double bond must change its place to be suitable for a
phenanthrene complex in **XXV**. This type is then retained in
XXVI. The p bands are little affected except between **XXV** and
XXVI when the new tetraphene-type appears.[2]

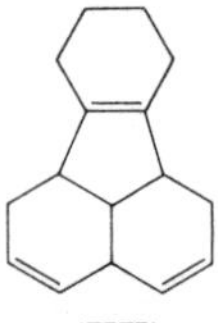

(XX)

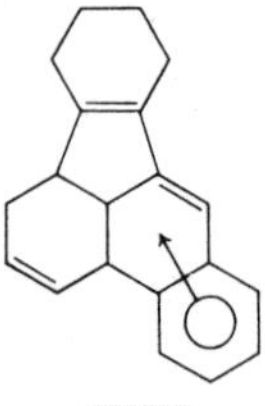

(XXV)

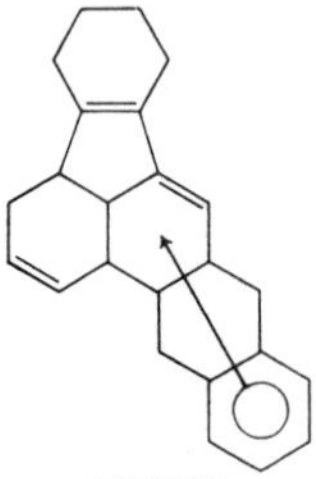

(XXVI)

$$\lambda_p = 3585 \xrightarrow{+105} 3690 \xrightarrow{-10} 3680 \text{ Å}$$
$$\lambda_\beta = 2370 \xrightarrow{+200} 2570 \xrightarrow{+610} 3180 \text{ Å}$$

The position of the double bonds in fluoranthene (**XX**) remains
unaltered in periflanthene (**XXVII**) which reacts with maleic anhydride and chloranil to form the anhydride **XXVIII**. Decarboxylation yields the benzologue **XXIX**. This involves the usual big
changes such as when going from perylene to benzoperylene, i.e.
a big shift of the p bands to shorter wavelength and a red shift
of the β bands.[3]

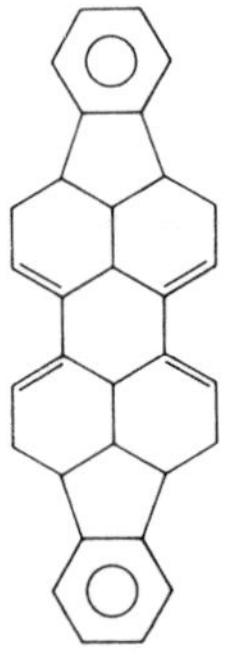

(XXVII) 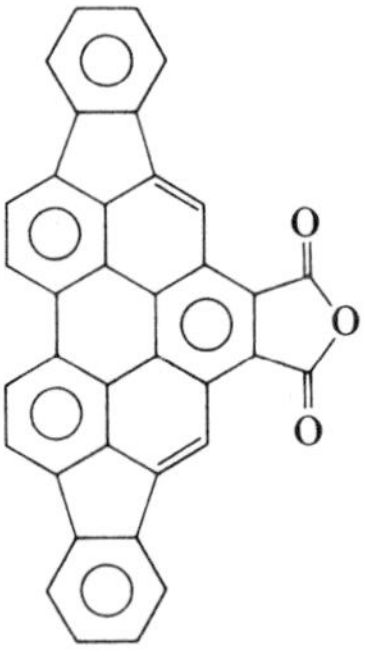   **(XXVIII)**    **(XXIX)**

$$\lambda_p = 5400 \xrightarrow{-650} 4750 \text{ Å}$$
$$\lambda_\beta = 2940 \xrightarrow{+210} 3150 \text{ Å}$$

The annellation of *o*-phenylene complexes to naphthalene shows some interesting features in the n.m.r. spectrum:

(In Hz ex TMS at 100 MHz in CS_2.)

The first phenylene complex causes a big shift towards lower field. This is particularly striking for the bay protons (+45.8 Hz). Simultaneously double-bond fixation is enforced by the five-membered ring in fluoranthene. Since a second five-membered ring in **IV** produces the opposite effect the original freedom of double-bond movement as in naphthalene is restored. Therefore big shifts toward higher field are recorded which make the bay effect disappear leaving only small shifts by comparison with naphthalene. This accounts for the formation of single bonds but not for ring currents in the five-membered rings in **IV**. The strain imposed by the five-membered rings causes the β,β position of the naphthalene complex in the CH_3 derivate **XXX** to assume some double-bond character. This becomes obvious in the coupling of the CH_3 group with the adjacent aromatic protons which causes the CH_3 signal to form a doublet with a separation of 0.4 Hz. This cannot be observed in other naphthalene derivatives.

(XXX)

The annellation of two naphthalene complexes to benzene does not produce a similar effect. The big shifts for the first annellation are maintained for the second annellation and even slightly increased as expected:

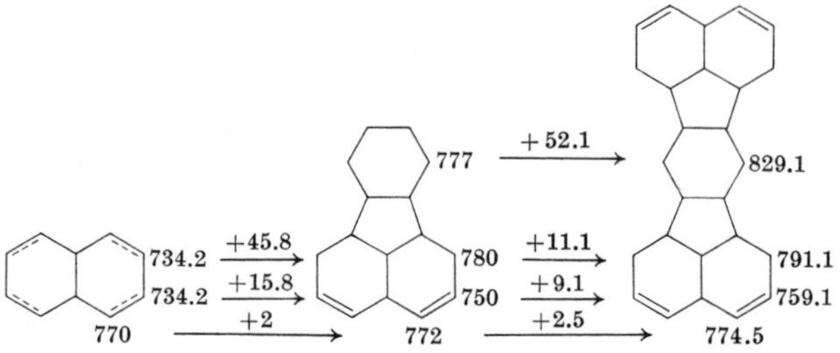

(XXXI)

It is very striking that the *peri*-phenylene ring in fluoranthene has two protons in the 7- and 10-positions which shift +50 Hz to lower field by comparison with benzene (727 Hz). This shift (+52.1 Hz) is repeated when passing to naphthylene-fluoranthene (**XXXI**).

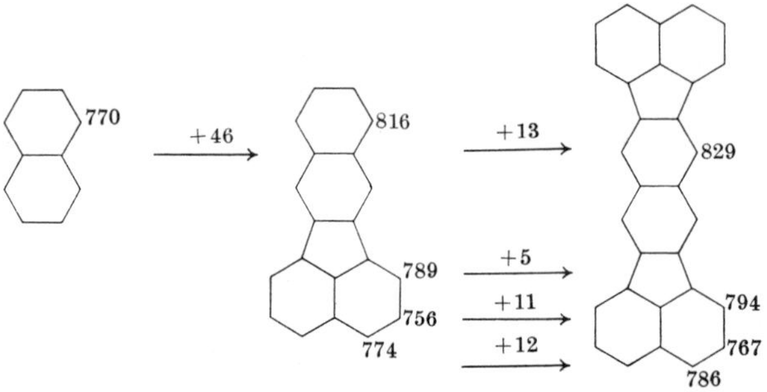

(XXIII)   **(XXXII)**

A very asymmetric annellation effect is recorded in passing from naphthalene to benzofluoroanthene (**XXIII**) and **XXXII**. The α protons of the central naphthalene complex shift by +46 and +13 Hz, respectively. This is in accordance with the electronic spectra (p. 97).

References

1. E. Clar and J. F. Stephen, *Tetrahedron*, **20**, 1559 (1964).
2. E. Clar, A. Mullen, and Ü. Sanigök, *Tetrahedron*, **25**, 5639 (1969).
3. K. F. Lang and M. Zander, *Chem. Ber.*, **95**, 673 (1962).

CHAPTER 22

Zethrene and Heptazethrene Series

The Zethrene Series

Hydrocarbons of the zethrene-type have fixed double bonds in the centre even if written with conventional Kekulé structures. The violet-red hydrocarbon zethrene (**I**) has the typical absorption spectrum of an aromatic hydrocarbon, obviously unimpaired by the formally fixed double bonds. The conclusion that double-bond fixation has no effect on the aromatic character of a spectrum has already been reached in the phene series where sextets enforce bond fixation. However, Kekulé structures do not.

The annellation effects in the zethrene series are similar to those in the perylene series. The extension of the naphthalene complex brings no increase in the number of sextets and big shifts in going to benzozethrene (**II**) and dibenzozethrene (**III**). The latter is so reactive that its heptachloro derivative had to be used for the spectrum. In contrast the dibenzozethrene **IV** has two more

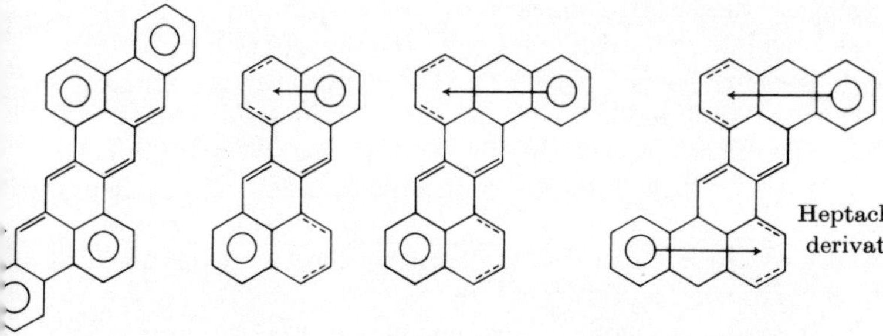

Heptachloro-derivative

(**IV**)   (**I**)   (**II**)   (**III**)

$$5350 \xleftarrow{\;-150\;} 5500 \xrightarrow{\;+610\;} 6110 \xrightarrow{\;+1110\;} 7220 \text{ Å}$$

$$3525 \xleftarrow{\;+445\;} 3080 \xrightarrow{\;+135\;} 3215 \xrightarrow{\;+445\;} 3660 \text{ Å}$$

sextets than zethrene. In accordance with this there is a small shift of the p bands to shorter wavelength and a big red shift of the β bands.

Whilst the benzologues **II** and **III** which are blue and green, respectively, are far more reactive and unstable than zethrene, the stabilizing influence in the red dibenzozethrene **IV** is very noticeable.[1]

The fixed double bonds in the zethrene series can be proved by the n.m.r. spectra. The bay protons of zethrene (**I**) are at 756 Hz for the protons on the linear branch and at 810 Hz (both from TMS at 100 MHz) for the protons on the naphthalene complex[2]. Since protons in the *meso* position usually absorb at lower field than protons in terminal rings, the reversed recording must be related to the fixed double bonds which belong to empty rings.

(I)   (IV)

This result is confirmed in dibenzozethrene (**IV**). Here again the protons on the linear system which belong to empty rings are at higher field.

The Heptazethrene Series

Heptazethrene (**V**) consists of two naphthalene complexes connected by four fixed double bonds. It is a green (violet in solution) hydrocarbon of great reactivity. The heptazethrene complex is very much stabilized by two more sextets in dibenzoheptazethrene (**VI**). The difference between these two hydrocarbons is similar to that between phenanthrene and naphthalene. Dibenzohepta-zethrene (**VI**) is deep green and gives violet solutions. In passing from **V** to **VI** the p bands shift slightly to shorter wavelength and the β bands considerably to the red.

Extending the naphthalene complexes to two anthracene complexes yields no new sextet and makes the green dibenzo-heptazethrene (**VII**) so unstable that it cannot be isolated in a pure state. These annellation effects are exactly as expected.

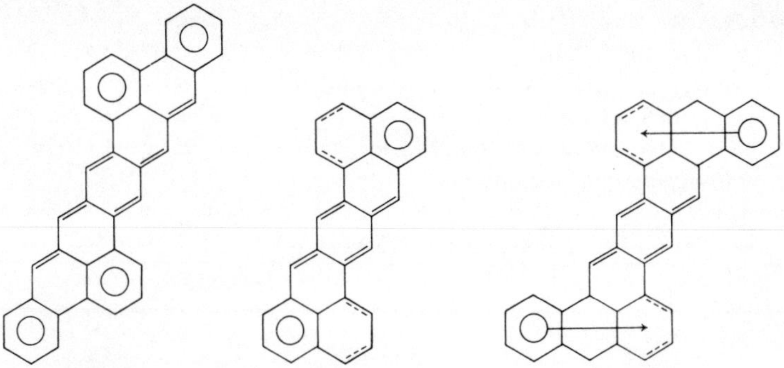

| (VI) | (V) | (VII) |

$\lambda_p = 5720 \xleftarrow{-140} 5860$ Å

$\lambda_\beta = 3640 \xleftarrow{+100} 3540$ Å $\xrightarrow{\hspace{2cm}}$ Green, unstable

An interesting property of the heptazethrenes is their basic character which enables them to form hydrochlorides **IX**. These must be derived from the polarized structure **VIII** which leaves the sextets intact but transforms the four fixed double bonds into a new sextet and a positive and a negative C atom. The proton addition takes place to the lone pair of electrons of the latter. The hydrochlorides are oxidizable and yield chloroheptazethrene (**X**). The dibenzoheptazethrene (**VI**) behaves in the same way.

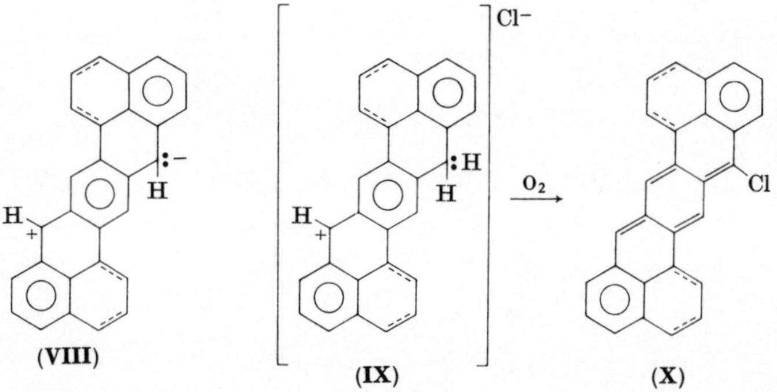

| (VIII) | (IX) | (X) |

The tendency to form polarized structures is certainly connected with the energy gain resulting from the formation of a new sextet in the centre of the molecule.[3]

References

1. E. Clar, K. F. Lang and H. Schulz-Kiesow, *Chem. Ber.*, **88**, 1520 (1955); E. Clar, I. A. Macpherson and H. Schulz-Kiesow, *Liebigs Ann.*, **669**, 44 (1963).
2. R. H. Mitchell and F. Sondheimer, *Tetrahedron*, **26**, 2141 (1970); and private communication.
3. E. Clar, G. S. Fell and M. H. Richmond, *Tetrahedron*, **9**, 96 (1960); E. Clar and I. A. Macpherson, *Tetrahedron*, **18**, 1711 (1962).

CHAPTER 23

Even-Alternant and Odd-Alternant Hydrocarbons

In 1932 Schmidt[1] introduced the alternant marking of aromatic hydrocarbons and gave it the meaning that C atoms with the same marking have π electrons with the same spin. Thus bonds can only be established between C atoms with different markings and π electrons with opposed spins. Izmailski[2] pointed out that the alternant formula for benzene (**I**) contains implicitly all the canonical structures **II, III, IV, V** and **VI** of the valence-bond theory. This is certainly a big advantage, because one would not have to assume these structures to let them disappear in a resonance hybrid.

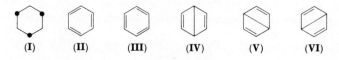

(**I**)   (**II**)   (**III**)   (**IV**)   (**V**)   (**VI**)

This spin distribution was rejected in 1938 by Hückel[3] who, for statistical reasons, wanted other spin distributions admitted, provided the number of marked and unmarked C atoms is equal.

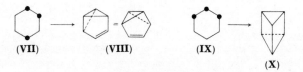

(**VII**)   (**VIII**)   (**IX**)

(**X**)

This proved later to be a very far-reaching proposal. It could mean that the spin distribution in **VII** would lead to the hydrocarbon benzvalene (**VIII**) which is certainly quite different from

107

benzene. The spin distribution in **IX** could yield the hydrocarbon prismane (**X**), which was later synthesized and is equally completely different from benzene.

A question which must now be asked is whether the only condition necessary is the acceptance of the alternant principle requisite for the stability of an aromatic hydrocarbon composed of hexagons and whether one can proceed with the arrangement of π electrons in orbitals occupied by pairs of π electrons with opposed spins. Here again only experiments can settle the question.

Let us consider the hydrocarbon triangulene $C_{22}H_{12}$ (**XI**). It has no Kekulé structure and cannot be formulated with sextets except as a diradical (**XII**) consisting of an anthracene and benzene complex. One could also formulate it with a frame of a cyclic polyene system and a centre consisting of four π electrons (marked with points) which might be symmetrically arranged in two orbitals, but not in double bonds (**XIII**). There are contradicting predictions about the stability of such a system.[4]

(**XI**)   (**XII**)   (**XIII**)

(**XIV**)

Outside the polyene frame this centre piece of four C atoms does exist. It is the hydrocarbon methylenecyclopropane (**XIV**) which is quite stable.[5] The cyclopropane ring would require a bond between *meta*-C atoms in triangulene. There are no polycyclic hydrocarbons known with a bond between *meta*-C atoms. Methylenecyclopropane can also exist as a diradical (**XV**).[6] This compound which tends to dimerize to **XVI** is rather more related to the centre system in triangulene, because it shows that there cannot be a bond between *meta*-C atoms.

(XV)   (XVI)

The experimental facts prove that triangulene is not an aromatic hydrocarbon of any stability but a diradical. The attempted synthesis yields hexahydrotriangulene (**XVII**) which has the u.v. spectrum of an alkyl pyrene. If its dehydrogenation is carried out under standard conditions, i.e. at 300° with a Pd–C catalyst, it disappears into the catalyst and cannot be recovered from it even in high vacuum at 500°. Therefore it must be concluded that triangulene is a diradical which polymerizes immediately after its formation. A dehydrogenation at lower temperature at 210° in trichlorobenzene with Pd–C gives the same result. A brown polymer is formed without any spectral evidence of the formation of triangulene.

The non-existence of the triangulene skeleton can be proved at even lower temperature and milder conditions. Triangulene-quinone (**XVIII**) is a deep red stable compound. If the triangulene system were unstable then triangulene quinone should not give a vat, i.e. a water-soluble disodium salt of the dihydroxy derivative (**XIX**), which would be a diradical like triangulene itself. If the quinone **XVIII** is reduced with aqueous alkaline sodium dithionite a deep-green solution is formed from which a green salt crystallizes. The analysis of this salt shows the presence of one atom of sodium in accordance with the normal formula **XX** and not a disodium salt as required by the diradical formula **XIX**.[7]

(XVII)   (XVIII)   (XIX)   (XX)

In spite of its instability triangulene (**XXI**) is an alternant hydrocarbon. There must be an additional element of stability in stable aromatic hydrocarbons. This can be found by a confrontation of triangulene (**XXI**) with its stable isomer, the yellow anthanthrene (**XXII**) (see p. 81). In triangulene there are ten marked and twelve unmarked C atoms. Therefore, there is an excess of two π electrons with the same spin. These cannot be paired in double bonds or otherwise. It is obvious that the alternant marking is the overriding principle. If it gives two or more spins in excess then the system is unstable. It can be shown that triangulene can be marked with an equal number of marked and unmarked C atoms, but not in an alternant way. The name "odd-alternant" is proposed for compounds of this type.

Marked C• = 10		Marked C• = 11
Unmarked C = 12		Unmarked C = 11
Difference = 2		Difference = 0

(**XXI**)   (**XXII**)

In anthanthrene there are eleven marked and eleven unmarked C atoms which give the system its stability. If the markings are equated with aromatic π-electron spins then anthanthrene must have a singlet ground state.

A similar comparison can be made between the stable zethrene **XXIII** and the hydrocarbon **XXIV**. Zethrene has twelve marked and twelve unmarked C atoms; therefore it must be stable. Moreover, the fixed double bonds do not make any difference. The isomeric hydrocarbon **XXIV** contains eleven marked and thirteen unmarked C atoms. There must be an excess of two π electrons with the same spin. In fact the hydrocarbon **XXIV** could not be synthesized because it is odd-alternant.

Marked C• = 12		Marked C• = 13
Unmarked C = 12		Unmarked C = 11
Difference = 0		Difference = 2

(**XXIII**)   (**XXIV**)

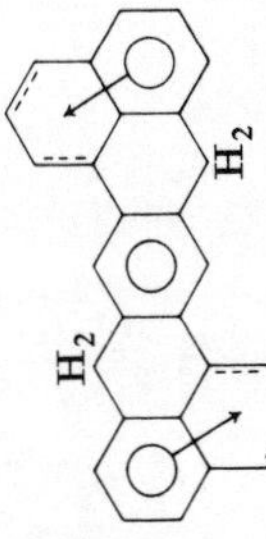

(XXV) →

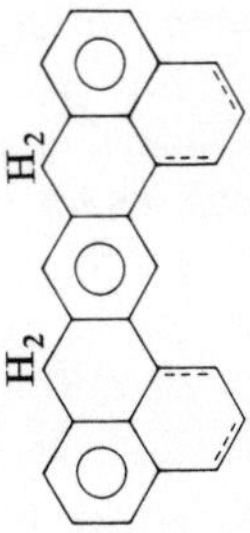

(XXVI) →

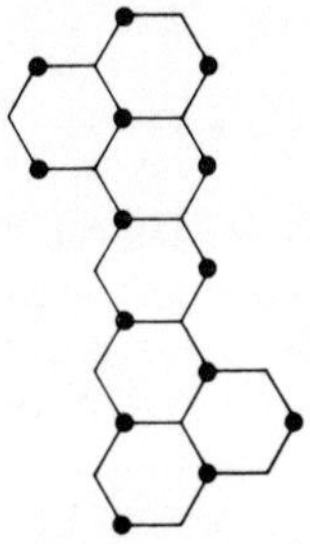

(XXVII)

Marked C• = 14
Unmarked C = 14
Difference = 0

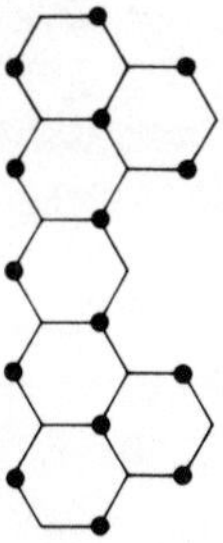

(XXVIII)

Marked C• = 15
Unmarked C = 13
Difference = 2

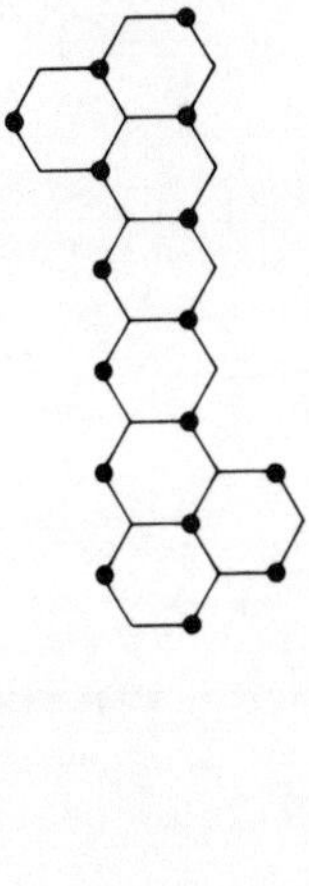

(XXIX)

Marked C• = 16
Unmarked C = 16
Difference = 0

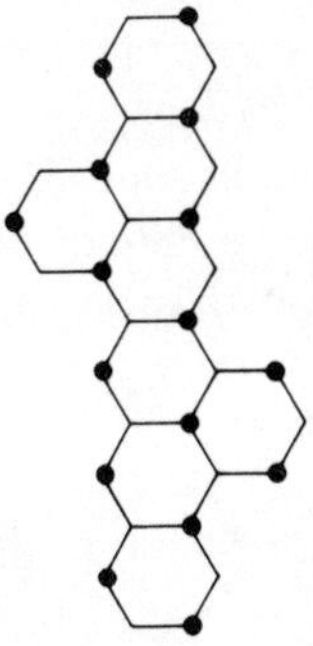

(XXX)

Marked C• = 16
Unmarked C = 16
Difference = 0

Dihydroheptazethrene (**XXV**) and dihydrodibenzopentacene (**XXVI**) are very similar hydrocarbons. However, they behave completely differently on dehydrogenation. Hydrocarbon **XXV** is dehydrogenated under the mildest conditions to heptazethrene (**XXVII**) whilst hydrocarbon **XXVI** needs rather drastic conditions and does not yield dibenzopentacene (**XXVIII**) but a polymer of it. The comparison of the markings shows that heptazethrene is an even-alternant and dibenzopentacene an odd-alternant hydrocarbon.[8]

Both the blue octazethrene (**XXIX**)[9] and the red dibenzozethrene (**XXX**)[10] can be prepared. Both are even-alternant hydrocarbons as the marking shows, but the very great difference in reactivity cannot be demonstrated in this way. This is a monopoly of the strict application of the aromatic sextet which gives octazethrene two and dibenzozethrene four sextets. The implications are obvious.

All hydrocarbons of triangular shape are odd-alternant and radicals. The first member of this series perinaphthyl or phenalyl is a monoradical, triangulene is a diradical, the next higher member **XXXI** is a triradical and the next a four-fold radical (**XXXII**). Their "oddity" does not consist in the number of odd electrons but in the remarkable fact that hydrocarbons built up of hexagons in a triangular shape must be radicals.

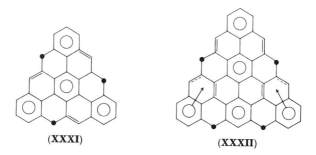

(**XXXI**)   (**XXXII**)

References

1. O. Schmidt, *Ber. Deut. Chem. Ges.*, **67**, 1870, 2078 (1934); **68**, 60, 356, 795, 1026 (1935).
2. V. A. Izmailski, *Dokl. Akad. Nauk S.S.S.R.*, **60**, 395 (1948); *Chem. Abstr.*, **1949**, 40.

3. E. Hückel, *Grundzuge der Theorie ungesättigter and aromatischer Verbindungen*, Verlag Chemie, Berlin, 1938, p. 28.

4. H. C. Longuett-Higgins, *J. Chem. Phys.*, **18**, 265 (1950); M. W. Lister, *Canad. J. Chem.*, **35**, 934 (1957).

5. J. T. Gragson, K. W. Greenlee, J. M. Derfer and C. E. Boord, *J. Amer. Chem. Soc.*, **75**, 3344 (1953).

6. P. S. Skell and R. J. Doerr, *J. Amer. Chem. Soc.*, **89**, 4688 (1967).

7. E. Clar and D. G. Stewart, *J. Amer. Chem. Soc.* **75**, 2667 (1953); **76**, 3504 (1954).

8. E. Clar and I. A. Macpherson, *Tetrahedron* **18**, 1411 (1962).

9. K. R. Erünlü, *Liebigs Ann.*, **721**, 43 (1969).

10. E. Clar, I. A. Macpherson and H. Schulz-Kiesow, *Liebigs Ann.*, **669**, 44 (1963).

CHAPTER 24

On the Possibility of *para* π Bonds

After Kekulé published his formula for benzene with the three double bonds Dewar proposed a formula with a long diagonal bond (**I**). This structure was adopted by the valence-bond theory as one of the contributing structures to the resonance hybrid. It is also implicit in the alternant marked formula for benzene p. 107.

(**I**)   (**II**)   (**III**)   (**IV**)   (**V**)

The Dewar formula gained new interest when Dewar benzene **II** was synthesized. Its n.m.r. spectrum shows that the diagonal bond is a σ bond because it has two protons bound to tertiary C atoms at 6.16 τ Hz. It is not planar, so that the possibility of a *para* π bond still remains open. It looks particularly attractive in anthracene (**III**) because the side rings could have benzenoid character. This structure is called the Anschütz formula. The latter thought he had proved it by a synthesis of anthracene from benzene, tetrachloroethane and aluminium chloride. Later X-ray measurement showed that the *para* bond is too long for a normal single bond.

Anschütz's anthracene formula was almost forgotten for many years until new facts came to light. It was the surprising result of the n.m.r. spectrum of methylanthracene (**IV**) that the CH_3 signal appeared as a doublet with a separation of 0.8 Hz. Decoupling of the *para* proton clearly showed that the spin–spin coupling originated from the *para* proton. It is known from 9-methyl-phenanthrene **V** that the CH_3 doublet has a separation of 1.0 Hz when coupling originates from the *ortho* proton. It is further known from 2,4,5,6-tetrachlorotoluene that the coupling constant of the benzylic *meta* coupling is 0.36 Hz.[1] If the magnetic information is transmitted through the C—C—C—C chain one would expect the *para* coupling of the CH_3 in anthracene to be even smaller, perhaps 0.1 Hz.

Therefore a coupling constant of 0.8 Hz in 9-methylanthracene suggests that there is a direct magnetic interaction between the *para*-C atoms in anthracene. This appeared to be confirmed by the fact that 6-methylanthanthrene (**VI**) showed no CH_3 doublet but a singlet in the n.m.r. spectrum. Here the distance between the 6- and 12-positions is too long. Moreover, the path is blocked by a single bond. Further confirmation came from the discovery of a CH_3 doublet (separation 1 Hz) in the n.m.r. spectrum of 5-methyl-tetracene (**VII**), 7-methyltetraphene (**VIII**) (0.8 Hz) and 12-methyltetraphene (**IX**) (0.8 Hz).

(**VI**)   (**VII**)   (**VIII**)   (**IX**)

The situation became confused when it was found that in 1-chloroanthracene (**X**) the non-equivalent *para* protons do not form a doublet; however, a doublet (0.8 Hz) was found in 1-chloro-10-methylanthracene (**XI**). There are also no doublets from *para* protons in 1.2,7.8-dibenzanthracene (**XII**), and in 5-bromotetracene (**XIII**) only a slight broadening of the singlets was found. This was also found in the singlets in 5,11-dibromotetracene (**XIV**), 5,12-dibromotetracene (**XV**), 1,4-dichlorotetracene (**XVI**) and in similar compounds and could be related to *peri* coupling. Moreover, a broadening of signals cannot be used as evidence for *para* coupling if the CH_3 signals show clear doublets produced by interaction along a longer path. It is even more puzzling that non-equivalent *para* protons in terminal rings (**XVII**) produced *para*

(X)   (XI)   (XII)   (XIII)

(XIV)   (XV)   (XVI)

coupling shown by clear signal splitting,[2] if this is not so in the
simpler singlets of *para* (*meso*) protons. Only a reinvestigation of
the n.m.r. spectra of pyridazine[3] and its CH_3-derivatives provided
a beginning to the solution of the problem. The coupling constants
of pyridazine show that the double bond is fixed in the 4,5-position
(**XVIII**). This is confirmed by the n.m.r. spectrum of 4-methyl-
pyridazine (**XIX**) which has a CH_3 doublet with a separation of
1.0 Hz. The coupling with the proton in the 5-position can be
proved by decoupling. 3,6-Dimethylpyridazine (**XX**) and 3-methyl-
6-chloropyridazine (**XXI**) show no CH_3 doublets but sharp singlets.
There can be no coupling with the neighbouring protons and there
cannot be adjacent double bonds. This being the case the question
now arises as to why the CH_3 group in 3-methylpyridazine (**XXII**)
produces a doublet with a separation of 0.5 Hz. In this case the
magnetic information would be passed on through a chain of four
C atoms whilst in **XX** and **XXI** no doublet appeared when the
path goes through only two C atoms of the ring.

The only answer is that there must be a direct *para* bond in
3-methylpyridazine (**XXII**) which is probably a π bond.[4] Since it
is indispensable to use a different symbol for a π bond to that for a

(XVII)   (XVIII)   (XIX)   (XX)   (XXI)   (XXII) σ bond it is proposed to mark the former with a dotted line as in **XXII**. This π bond could be weak in pyridazine and anthracene but would become much stronger through hyperconjugation of the CH₃ group in 3-methylpyridazine and 9-methylanthracene. Hyperconjugation appears to be a much stronger factor than assumed so far. It does explain why the weak *para* bond does not produce measurable doublets in non-equivalent *meso*-protons but gives clear doublets in methyl derivatives. It should be recalled that the CH₃ group in methylanthracene has a particular character and gives reactions not found in other CH₃ derivatives. For instance it yields bromomethylanthracene (**XXIII**) instead of the expected 10-bromo-9-methylanthracene.

One could imagine that the weak *para* bond is the reason for the reaction of anthracene and the higher acenes with maleic anhydride under mild conditions (**XXIV**→**XXV**). This *para* bond may also be present in diphenylbenzofurane (**XXVI**) which reacts very readily with maleic anhydride.

The weak *para* bond could also be present in dibenzoperylene (**XXVII**) where it could allow the formation of an induced sextet in the centre. This would account for the low-field protons in the bay position (854 and 833 Hz) which cannot be explained by formula **XXVIII** which has an empty centre "E" like perylene (see p. 61). It is no coincidence that dibenzoperylene (**XXVII**) reacts very readily with maleic anhydride.

Tetrabenzoperylene (**XXIX**) reacts very readily twice with maleic anhydride to form the dianhydride **XXXI**. This reaction is difficult to explain without the weak *para* bond in **XXIX**. After the first addition another *para* bond can be formed in **XXX** which leads to the second addition. The dianhydride **XXXI** gives coronene-tetracarboxylic dianhydride on dehydrogenation. It is difficult to explain the reaction by formula **XXXII** which has an empty centre "E" like perylene and does not account for the low-field bay protons at 872 Hz from TMS.

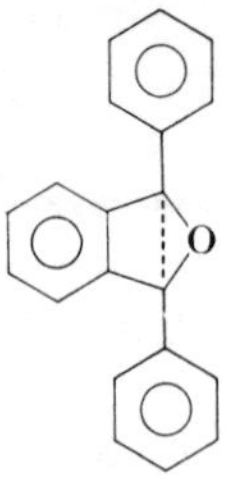

(XXVI) 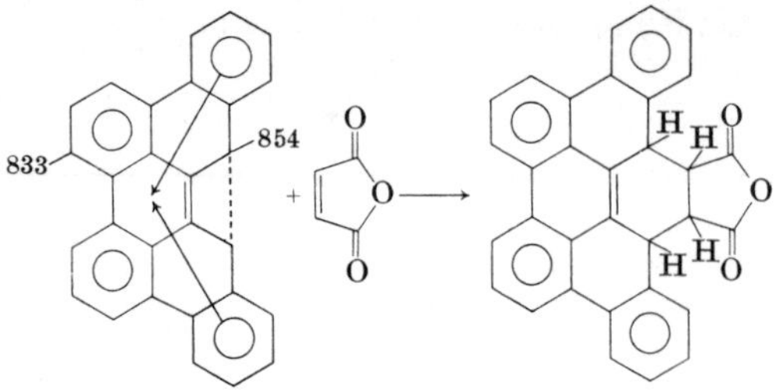   (XXVII)

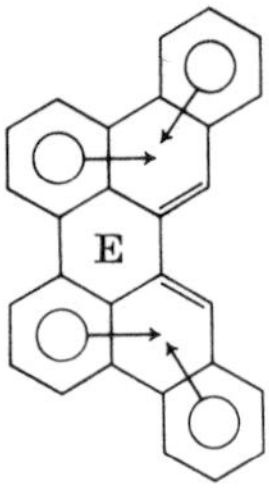

(XXVIII)

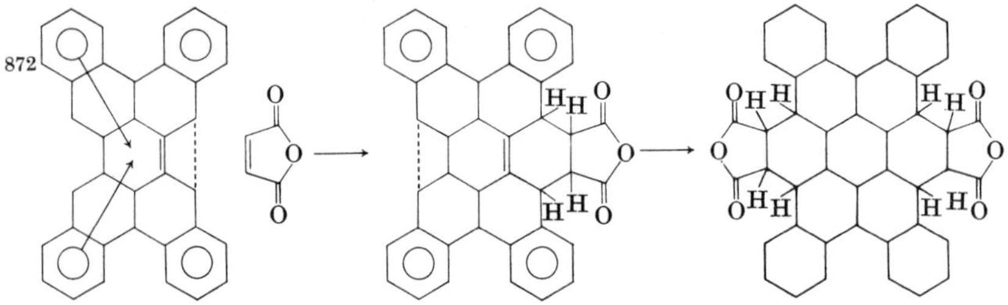

(XXIX)   (XXX)   (XXXI)

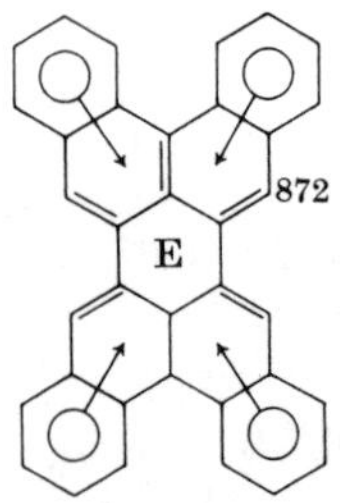

(XXXII) 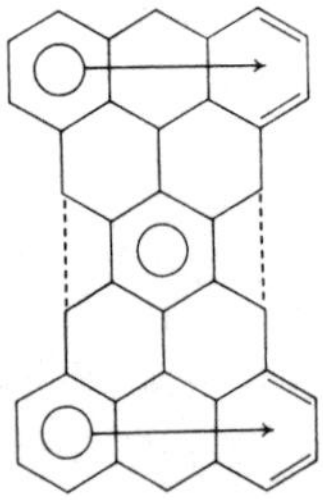   (XXXIII) 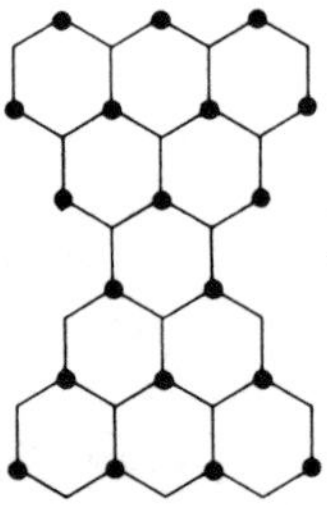   (XXXIV)

Marked C• =
Unmarked C =
Difference =

The whole problem of the *para* π bond might appear in a new light if the following facts are taken into account. The CH_3 doublet of 3-methylpyridazine (**XXII**) disappears on heating and is replaced by a broad singlet which remains if the compound is then cooled in liquid nitrogen. Kept in the frozen state only a very slow return of the CH_3 doublet is observed; this takes weeks. A quicker conversion takes place in CS_2 solution at room temperature within days.

This change is accompanied by significant changes in the u.v. spectrum. The broad absorption band at 3420 Å of 3-methylpyridazine is obviously associated with the *para* π form (**XXII**) which produces the CH_3 doublet because it is weakened on heating and partly replaced by a group of sharp benzene-like bands at 3080, 3000, 2930, 2860 and 2800 Å. The latter group of bands appears to belong to the benzenoid form of 3-methylpyridazine without a *para* π bond. The benzene-like group of sharp bands disappears first quickly then more slowly at room temperature in cyclohexane solution.

The conclusion from these observations could be far-reaching. If the above changes could be observed in other polycyclic compounds with α-CH_3 groups in terminal rings then the separate existence of Dewar π structures must be admitted. These would not participate in a resonance hybrid of fictitious Kekulé structures but would have a separate existence of variable stability. It is striking that just these compounds have α-CH_3 signals which are asymmetric multiplets.[5]

It must be admitted that the use of the weak *para* bond (symbolized by a dotted line) is just in its beginning. The idea will stand on more solid ground when the hydrocarbon **XXXIII** has been synthesized and found to be stable. It consists of two condensed triangulene systems and has two *para* bonds. It cannot be written with Kekulé structures but the marking in **XXXIV** shows that it is an even-alternant hydrocarbon which should exist and easily react twice with maleic anhydride.

The *para* π bond may not be so weak after all. One has to consider that there is not a σ bond under the π bond as in a double bond. This would cause the π electron to receive more nuclear charge and the bond would become much stronger. The increase of bonding would be compensated by the greater distance between the C atoms. It could well be that the *para* structure of naphthalene (**XXXV**) is just a little less energetically favoured than structure **XXXVI**. This could account for the fact that naphthalene reacts with maleic anhydride with a yield of only 1 per cent.

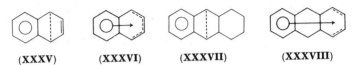

(**XXXV**)   (**XXXVI**)   (**XXXVII**)   (**XXXVIII**)

In anthracene the *para* structure (**XXXVII**) may be equal to or more favourable than structure **XXXVIII**, which would result in a complete reaction with maleic anhydride. The tendency to form *para* structures is probably predominating in the higher acenes. This is the reason why middle rings have not been marked with double bonds as is shown in **XXXVIII**.

References

1. H. Rottendorf and S. Sternhell, *Australian J. Chem.*, **17**, 1315 (1964).
2. C. W. Haig and R. B. Mallion, *Molec. Phys.*, **18**, 737 (1970).
3. C. C. Mackay, Thesis, Glasgow, 1971, p. 25; K. Tari and M. Ogata, *Chem. Pharm. Bull.* (*Japan*), **12**, 292 (1969).
4. E. Clar and C. C. Mackay, Future communication.
5. E. Clar and C. C. Mackay, *Tetrahedron*, **28,** in press (1972).

Index

121